Synthesis Lectures on Engineering, Science, and Technology

The focus of this series is general topics, and applications about, and for, engineers and scientists on a wide array of applications, methods and advances. Most titles cover subjects such as professional development, education, and study skills, as well as basic introductory undergraduate material and other topics appropriate for a broader and less technical audience.

Khalid Shaikh · Rohit Thanki · Sejal Shah

Artificial Intelligence in Genomics

Methods, Applications, and Clinical Translation

Springer

Khalid Shaikh
Prognica Labs
Dubai, United Arab Emirates

Rohit Thanki
Senior IEEE Member
Rajkot, Gujarat, India

Sejal Shah
Department of Bioinformatics
Marwadi University
Rajkot, Gujarat, India

ISSN 2690-0300 ISSN 2690-0327 (electronic)
Synthesis Lectures on Engineering, Science, and Technology
ISBN 978-3-032-25485-6 ISBN 978-3-032-25486-3 (eBook)
https://doi.org/10.1007/978-3-032-25486-3

This Springer imprint is published by the registered company Springer Nature Switzerland AG
The registered company address is: Gewerbestrasse 11, 6330 Cham, Switzerland

Preface

The field of genomics has undergone a transformative evolution, driven by the convergence of high-throughput sequencing technologies and artificial intelligence (AI). This book explores the intersection of these two domains, offering a comprehensive overview of how AI is reshaping genomic research, precision medicine, and drug discovery. From foundational concepts to cutting-edge applications, the chapters delve into the technical, ethical, and regulatory dimensions of genomic AI.

Chapter 1 introduces the shift from genetics to genomics, emphasizing the multifactorial nature of complex diseases and the indispensable role of AI in analyzing vast genomic datasets. It lays the groundwork for understanding the central dogma of molecular biology and the complexities of gene regulation, which are critical for applying AI effectively in genomics.

Chapter 2 focuses on the computational foundations of genomic data, detailing the types of data, file formats, and preprocessing techniques required for accurate analysis. It also explores feature representation, high-performance computing, and scalable genomic data pipelines, providing the technical bedrock for modern genomic research.

Chapter 3 examines AI methodologies in genomic analysis, progressing from classical machine learning approaches to deep learning, transformer-based foundation models, and generative AI. It highlights the strengths, limitations, and applications of these methods while addressing challenges related to interpretability, data governance, and scalable computation.

Chapter 4 explores the applications of AI across the genomic analysis pipeline, from genome assembly and variant calling to gene expression modeling, epigenomic regulation, disease risk prediction, microbiome analysis, and population genomics. It emphasizes the importance of responsible AI deployment and biological interpretability.

Chapter 5 delves into the transformative role of AI in precision medicine and drug discovery. It covers multi-omics integration, patient stratification, biomarker discovery, molecular property prediction, and generative chemistry, showcasing how AI bridges the gap between biological data and therapeutic intervention.

Chapter 6 addresses the ethical, privacy, and regulatory considerations of genomic AI. It explores privacy-preserving technologies, algorithmic bias, explainability, and global standards, emphasizing the need for inclusivity, transparency, and accountability in genomic AI applications.

Chapter 7 looks ahead to the future of genomic AI, highlighting emerging opportunities such as next-generation foundation models, generative biology, digital twins, real-time edge genomics, and self-driving laboratories. It also discusses the ethical and biosecurity challenges posed by these advancements.

This book aims to provide researchers, clinicians, and policymakers with a structured understanding of how AI is revolutionizing genomics. By bridging technical innovation with ethical and regulatory frameworks, it seeks to ensure that genomic AI fulfills its promise of advancing precision medicine and improving global health.

Dubai, UAE — Khalid Shaikh
Rajkot, India — Dr. Rohit Thanki
Rajkot, India — Dr. Sejal Shah

Acknowledgments Our task has been made easier and the final version of this book considerably improved thanks to the support we have received. We extend our heartfelt thanks to the publishing team for their invaluable guidance and encouragement during the creation of this book. We also wish to thank our colleagues and peers for their insightful discussions and contributions that enriched the content.

Competing Interests The authors have no competing interests to declare that are relevant to the content of this manuscript.

Contents

List of Figures

List of Tables

1 Introduction to Genomics and Artificial Intelligence

1.1 Defining the Landscape: From Gene to Genomics

The journey from gene to genomics represents one of the most significant transformations in biological sciences. A gene is the fundamental unit of heredity, consisting of a specific sequence of DNA that encodes functional products such as proteins or RNA molecules. The molecular understanding of genes began with the discovery of the double-helical structure of DNA by Watson and Crick [1], which explained how genetic information is stored and replicated.

The development of molecular biology techniques enabled scientists to isolate and study individual genes. The invention of the Polymerase Chain Reaction (PCR) by Mullis and Faloona [2] revolutionized genetic research by allowing rapid amplification of specific DNA sequences. Similarly, the chain-termination method developed by Sanger, Nicklen, and Coulson [3] provided a reliable technique for DNA sequencing. These advances laid the foundation for identifying mutations, studying gene expression, and producing recombinant proteins for therapeutic purposes.

Despite these achievements, studying genes individually proved insufficient to explain complex biological systems. Many diseases, including cancer and diabetes, are polygenic and involve interactions among multiple genes and environmental factors. This limitation led to the emergence of genomics, defined as the comprehensive study of the entire genome of an organism.

A landmark achievement in genomics was the completion of the Human Genome Project, an international research initiative launched in 1990. The publication of the draft human genome sequence by Lander et al. [4] and Venter et al. [5] marked a historic milestone in science. Surprisingly, the human genome was found to contain approximately 20,000–25,000 protein-coding genes, far fewer than previously estimated. This discovery

K. Shaikh et al., *Artificial Intelligence in Genomics*, Synthesis Lectures on Engineering, Science, and Technology, https://doi.org/10.1007/978-3-032-25486-3_1

emphasized the importance of gene regulation, alternative splicing, and non-coding DNA in determining biological complexity.

The rapid evolution of sequencing technologies further accelerated genomic research. Next-Generation Sequencing (NGS) technologies allow simultaneous sequencing of millions of DNA fragments, making whole-genome sequencing faster and more cost-effective than traditional Sanger sequencing [6]. These technological advancements have given rise to various branches of genomics, including structural genomics, functional genomics, comparative genomics, and metagenomics.

Functional genomics focuses on understanding gene expression patterns using transcriptomics and proteomics. Comparative genomics analyzes similarities and differences among species to trace evolutionary relationships. Metagenomics enables the study of genetic material directly from environmental samples, significantly advancing microbiology and infectious disease research.

In medicine, genomics has led to the era of personalized or precision medicine, where diagnosis and treatment are tailored according to an individual's genetic profile. Genome-wide association studies (GWAS) help identify genetic variants associated with complex diseases. In oncology, genomic profiling assists in identifying driver mutations and selecting targeted therapies. Similarly, in agriculture, genomics accelerates crop improvement and enhances resistance to environmental stress. Our recent studies on personalize medicine targeting from candidate gene approach to next generation sequencing number of genes and pathways related to oral cancer [7–13].

In conclusion, the transition from gene-focused research to genomics reflects the evolution from reductionist biology to systems biology. While genes remain the basic units of heredity, genomics provides a holistic understanding of genetic networks and interactions. This shift continues to transform medicine, biotechnology, and life sciences in the 21st century.

The intellectual trajectory of modern biology can be characterized as a migration from the reductionist study of isolated components to the integrative analysis of complex systems. This paradigm shift is most clearly articulated in the distinction between genetics and genomics, two disciplines that, while etymologically linked and colloquially conflated, represent fundamentally different approaches to biological inquiry [14]. Understanding this distinction is not merely an exercise in semantics; it is a prerequisite for appreciating why artificial intelligence (AI) has become an indispensable tool in life sciences. Genetics provided the vocabulary of heredity, but genomics provided the "big data" substrate that necessitated the intervention of machine learning.

Genetics, the older of the two disciplines, is historically rooted in the study of heredity and the function of single genes [14]. Emerging from the Mendelian tradition, genetics focuses on the specific units of inheritance, genes and their transmission from parents to offspring. The primary investigative scope of genetics is often linear and causal, seeking to link a specific mutation in a specific gene to a specific phenotypic trait or disease state [15]. For example, the study of how a mutation in the CFTR gene leads to Cystic Fibrosis is a quintessential genetic inquiry. The methodology is typically targeted: researchers isolate a segment of DNA, amplify it, and analyze its function in isolation or within a controlled family lineage [16]. The National Human Genome Research Institute (NHGRI) defines genetics as the study of genes and their roles in inheritance, the way that certain traits or conditions are passed down from one generation to another [2].

Genomics, in contrast, is the comprehensive study of an organism's entire genetic makeup—the genome [14]. The term, coined more recently to reflect the scaling of biological inquiry, encompasses the sequencing, mapping, and analysis of all an individual's genes, including the complex interactions between those genes and, crucially, their interplay with the environment [15]. If genetics is the study of the "parts list" of a biological machine, genomics is the study of the "operating system" that governs how those parts interact to produce a living organism. The scope of genomics is inherently multifactorial and systemic. It does not ask "which gene causes this disease" but rather "how does the perturbation of this gene network, influenced by environmental factors, shift the system from a healthy state to a diseased state?" [14].

1.1.1 The Shift to Multifactorial Analysis

The necessity for genomics and subsequently for AI arises from the realization that the most burdensome diseases facing humanity are not monogenic but complex and multifactorial [14]. Conditions such as heart diseases, asthma, diabetes, and cancer are rarely the result of a single error in the genetic code. Instead, they arise from the subtle, non-linear interplay of hundreds or thousands of genetic variants (polygenic risk), each contributing a minute effect, combined with environmental triggers like diet, stress, and pollutant exposure [15].

Research institutions like the Jackson Laboratory highlighted that while genetics might involve experimental manipulation of specific loci in model organisms, genomics involves the computational analysis of vast human datasets to discover precise solutions for these complex diseases [16]. Researchers are discovering that it is virtually impossible to study a single gene without also studying other genes and the environmental factors that influence their expression [17]. This interconnectedness makes the two branches of molecular biology virtually synonymous in advanced research contexts, yet distinct in their analytical demands.

This holistic view transforms biology into a data science. The human genome contains approximately 3 billion base pairs. While a geneticist might manually analyze a sequence of a few thousand bases, a geneticist must analyze the entire sequence, including the extensive non-coding regions that regulate gene expression [15]. The sheer volume of this data exceeds human cognitive capacity. A human mind cannot intuitively grasp the correlation between 500,000 single nucleotide polymorphisms (SNPs) and the probability of developing Type 2 Diabetes. This cognitive bottleneck creates a niche for artificial intelligence. AI does not merely automate the analysis; it perceives patterns in high-dimensional space correlating millions of markers with phenotypic outcomes that traditional statistical methods might miss or oversimplify [19]. Thus, the shift from genetics to genomics is the foundational driver for the integration of machine learning in biology.

1.2 The Biological Substrate: Central Dogma and Regulatory Complexity

To apply artificial intelligence effectively to genomics, one must understand the "grammar" of the biological language it seeks to interpret. The foundational framework for this is the central dogma of molecular biology, a concept first articulated by Francis Crick in 1957 and published in 1958 [20, 21].

1.2.1 The Flow of Information

The central dogma described the directional flow of genetic information within a biological system, dictating how the static information stored in the genome is converted into the functional machinery of life. It is classically summarized as the pathway: **DNA → RNA → Protein** [21] (as shown). This pathway involves three primary processes:

- **Replication (DNA → RNA)**: This is the process by which existing DNA makes new DNA. It occurs during cell division, ensuring that the genetic information is accurately perpetuated and passed on to daughter cells [20]. The fidelity of this process is paramount; errors here become permanent mutations in the genome.
- **Transcription (DNA → RNA)**: This is the first step of gene expression. A specific segment of DNA (a gene) is read by an enzyme complex (RNA polymerase) and transcribed into a single-stranded sequence of RNA, specifically messenger RNA (mRNA) [22]. In eukaryotes, this process takes place inside the nucleus. The DNA sequence serves as a template, where nucleotide language of Adenine (A), Cytosine (C), Guanine (G), and Thymine (T) is transcribed into RNA, with Uracil (U) replacing Thymine [23]. The resulting mRNA is a "portable message" that carries the genetic instructions out of the nucleus [22].
- **Post-Transcriptional**: Gene expression in eukaryotic cells is a tightly regulated, multi-step process that extends beyond transcription and translation. Two critical regulatory stages are post-transcriptional modification (PTM at the RNA level) and post-translational modification (PTM at the protein level). These processes ensure accurate gene expression, functional diversity, and cellular adaptability. Post-transcriptional modification refers to molecular changes that occur to RNA after transcription but

before translation. In eukaryotes, the primary transcript (pre-mRNA) undergoes several essential modifications in the nucleus. One of the first steps is 5′ capping, in which a 7-methylguanosine cap is added to the 5′ end of the mRNA. This cap protects mRNA from exonuclease degradation and facilitates ribosomal recognition during translation [24]. Another critical modification is 3′ polyadenylation, where a poly(A) tail is added to the 3′ end. The poly(A) tail enhances mRNA stability and regulates nuclear export and translation efficiency [25]. RNA splicing is another major post-transcriptional event in which non-coding introns are removed, and coding exons are joined together. This process is carried out by the spliceosome complex and is essential for producing functional mRNA [26]. Alternative splicing further increases protein diversity by allowing a single gene to produce multiple mRNA variants, significantly expanding proteomic complexity in higher organisms [24]. Additionally, RNA editing, involving nucleotide modifications such as adenosine-to-inosine conversion, can alter the coding sequence and functional output of proteins [25]. Together, these modifications regulate mRNA stability, localization, and translational efficiency.

- **Translation (RNA → Protein)**: The mRNA molecule travels to the cytoplasm, where it binds to a ribosome, the cell's protein factory. Here, the genetic code is translated into the language of proteins. The ribosome reads the mRNA sequence in triplets called codons. Each codon specifies a particular amino acid. Transfer RNA (tRNA) molecules act as adapters, bringing the corresponding amino acids to the ribosome, which assembles them into a polypeptide chain [9]. This chain then folds into a functional 3D protein structure, which performs the work of the cell.
- **Post-Translational**: Post-translational modification, in contrast, refers to covalent and enzymatic changes that occur to proteins after they are synthesized on ribosomes. These modifications are crucial for protein activation, structural stability, localization, and degradation. One of the most common modifications is phosphorylation, the reversible addition of a phosphate group to serine, threonine, or tyrosine residues by kinases. Phosphorylation plays a central role in cell signaling and metabolic regulation [26]. Glycosylation, the attachment of carbohydrate moieties to proteins, occurs primarily in the endoplasmic reticulum and Golgi apparatus and is essential for protein folding, stability, and cell-cell recognition [24]. Ubiquitination involves the attachment of ubiquitin molecules to target proteins, marking them for proteasomal degradation and maintaining protein quality control [25]. Other modifications such as acetylation, methylation, disulfide bond formation, and proteolytic cleavage further refine protein function and cellular regulation.

A visualization of this process is shown in Fig. 1.1, depicting the double helix of DNA unwinding in the nucleus, the RNA polymerase creating a single-stranded mRNA transcript, and the mRNA exiting the nuclear pore to meet the ribosome. The figure should highlight the "flow" of information, perhaps using arrows to indicate the unidirectional

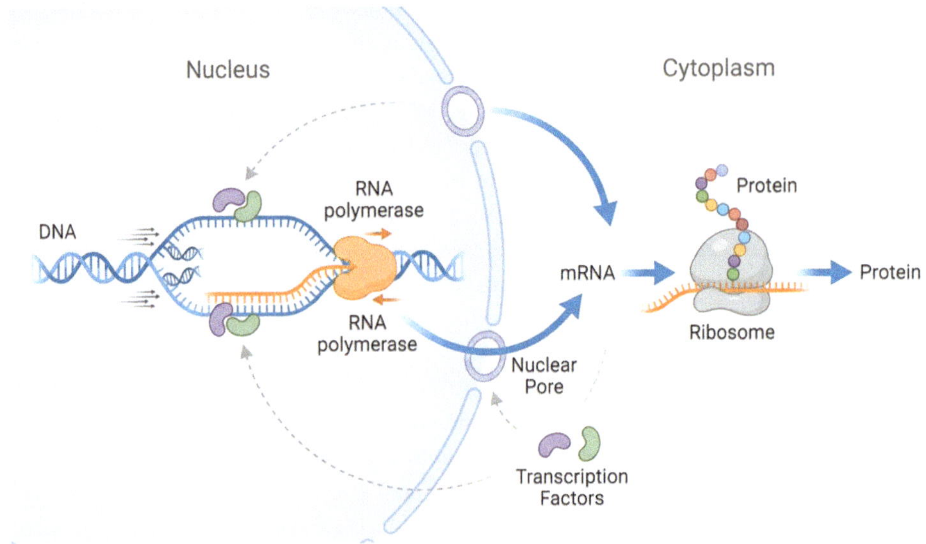

Fig. 1.1 Central dogma interaction map

nature of the canonical dogma, while subtly indicating feedback loops where proteins (transcription factors) return to the DNA to regulate the process.

1.2.2 Gene Architecture: Introns, Exons, and Splicing

While central dogma provides the basic schematic, the reality of eukaryotic gene structure adds layers of complexity that require sophisticated computational modeling. A gene is not a continuous, uninterrupted string of coding instructions. It is a mosaic of coding and non-coding sequences [27].

- **Exons**: These are the coding sequences that are retained in the final, mature mRNA and are ultimately translated into protein. They contain the actual "recipe" for the amino acid sequence [27].
- **Introns**: These are intervening, non-coding sequences that interrupt the exons. They are transcribed into the initial RNA molecule (pre-mRNA) but must be physically removed before translation can occur [27].

Introns are non-coding sequences within eukaryotic genes that are transcribed into pre-cursor messenger RNA (pre-mRNA) but are removed during RNA splicing. Although once considered "junk DNA," introns are now recognized as critical regulatory elements that influence gene expression, mRNA processing, and genome evolution. They contain important cis-regulatory sequences such as enhancers, silencers, insulators, and splicing regulatory elements that control transcriptional and post-transcriptional processes [24]. Structurally, introns are characterized by conserved splice junctions. Most eukaryotic introns follow the GU–AG rule, containing a 5′ splice donor site (GU), a 3′ splice acceptor site (AG), a branch point sequence with a conserved adenine, and a polypyrimidine tract. These elements are recognized by the spliceosome, a ribonucleoprotein complex composed of small nuclear RNAs (snRNAs) and associated proteins, which removes introns via a lariat intermediate during RNA splicing [28]. Proper recognition of these sequences is essential for accurate mRNA maturation.

One of the most significant regulatory features within introns is the presence of enhancers. Enhancers are DNA elements that increase transcriptional activity by serving as binding platforms for transcription factors. While enhancers are often located upstream or downstream of genes, many reside within introns. Intronic enhancers can regulate tissue-specific or developmental gene expression by facilitating chromatin looping, allowing transcription factors to interact with promoters over long genomic distances [29]. For example, intronic enhancers have been identified in genes involved in immune regulation and developmental pathways. Conversely, introns also contain silencers, which suppress gene transcription. Silencers function by binding transcriptional repressors that inhibit RNA polymerase II activity or alter chromatin structure to a repressive state. These elements ensure that genes are expressed only in appropriate cellular contexts and prevent aberrant transcription [30]. The balance between intronic enhancers and silencers contributes to precise spatiotemporal gene regulation.

In addition to transcriptional control, introns play a crucial role in alternative splicing. They harbor intronic splicing enhancers (ISEs) and intronic splicing silencers (ISSs), which regulate exon inclusion or exclusion. These elements interact with splicing factors such as SR proteins and heterogeneous nuclear ribonucleoproteins (hnRNPs), influencing splice site selection [28]. Alternative splicing significantly increases proteomic diversity, enabling a single gene to produce multiple protein isoforms. Introns may also contain genes encoding non-coding RNAs, including microRNAs (miRNAs) and long non-coding RNAs (lncRNAs). Intronic miRNAs are processed from host gene transcripts and participate in post-transcriptional gene silencing by targeting complementary mRNAs [31]. This demonstrates that introns contribute not only to gene regulation at the DNA level but also at the RNA level. Clinically, mutations within intronic regulatory elements can lead to disease. Alterations in splice sites or deep intronic mutations may result in aberrant splicing and defective protein products, as observed in disorders such as β-thalassemia and certain cancers [32]. Similarly, mutations affecting intronic enhancers or silencers can disrupt normal gene expression patterns.

The process of removing introns and joining exons is called splicing. This mechanism allows for alternative splicing, a phenomenon where a single gene can code for multiple different proteins (isoforms) by including or excluding specific exons [19]. This combinatorial capacity explains how the human genome, with only ~20,000 genes, can produce hundreds of thousands of distinct proteins. However, it also introduces a massive prediction challenge: determining which isoforms will be produced in which tissue type. Predicting splicing patterns from DNA sequences is a classic problem where deep learning models have demonstrated superior performance over traditional probabilistic models [19].

1.2.3 The Regulatory Landscape: Promoters and Enhancers

Perhaps the most significant challenge in genomics and the primary target for modern AI models like Enformer and Basenji is understanding gene regulation. Every cell in the human body contains the same DNA, yet a neuron is functionally and structurally distinct from a liver cell. This differentiation is governed by precise temporal and spatial control of gene expression [33]. This regulation is orchestrated by non-coding DNA elements:

- **Promoters**: These are DNA sequences located immediately upstream of a gene's start site. They serve as the docking platform for the basic transcription machinery (RNA polymerase and general transcription factors) and are responsible for initiating transcription [27].
- **Enhancers**: These are short regulatory regions (50–1500 bp) that can significantly increase the likelihood of transcription [34]. Unlike promoters, enhancers can be located up to 1 million base pairs (1 Mbp) away from the gene they regulate, positioned either upstream or downstream. They function in 3D space, looping into the chromatin to bring specific transcription factors into physical contact with the promoter [34].

There are hundreds of thousands of enhancers in the human genome, far outnumbering the protein-coding genes [34]. Identifying which enhancer regulates which gene is a massive combinatorial problem. Because enhancers do not have a fixed position relative to the gene, simple sequence scanning is insufficient. This is where AI becomes critical. Models like Enformer utilize transformer architectures to analyze long stretches of DNA sequence (up to 100 kb or more) to predict these 3D interactions and gene expression levels based on the presence of distal regulatory elements [35].

The non-coding regions, once dismissed as "junk DNA," are now understood to contain the majority of disease-associated variants identified by Genome-Wide Association Studies (GWAS) [33]. A mutation in an enhancer can break the "switch" that turns a gene on, causing disease even if the gene itself is intact. AI's ability to decipher this "cis-regulatory code" is unlocking the mechanisms of diseases that are not caused by broken proteins, but by broken regulatory circuitry [33].

1.2.4 Genomics to Epigenomics

The transition from genomics to epigenomics represents a major conceptual shift in molecular biology from understanding the static DNA sequence to deciphering the dynamic regulatory mechanisms that control gene expression without altering the nucleotide sequence itself. Together, genomics and epigenomics provide a comprehensive framework for understanding development, disease, and cellular identity. Genomics refers to the study of the complete set of DNAs within an organism, known as the genome. It encompasses gene sequencing, mapping, structural variation analysis, and functional annotation. The completion of the Human Genome Project marked a milestone in genomics, providing the first reference human genome and enabling large-scale identification of genes and regulatory regions [36]. Advances in next-generation sequencing (NGS) technologies have further accelerated whole-genome sequencing, comparative genomics, and genome-wide association studies (GWAS), allowing researchers to link genetic variants to complex diseases [37].

While genomics reveals what genes are present, it does not fully explain how identical DNA sequences can give rise to diverse cell types such as neurons, hepatocytes, or muscle cells. This gap led to the emergence of epigenomics, the study of genome-wide epigenetic modifications that regulate gene activity without changing the DNA sequence. Epigenetic mechanisms include DNA methylation, histone modifications, chromatin remodelling, and non-coding RNA-mediated regulation [38]. One of the most well-characterized epigenetic modifications is DNA methylation, typically occurring at cytosine residues in CpG dinucleotides. DNA methylation is associated with transcriptional repression and plays critical roles in genomic imprinting, X-chromosome inactivation, and suppression of transposable elements (Jones, 2012). Abnormal methylation patterns have been linked to cancers and other complex disorders.

Another central component of epigenomics is histone modification. DNA is wrapped around histone proteins to form nucleosomes, and chemical modifications such as acetylation, methylation, phosphorylation, and ubiquitination occur on histone tails. These modifications influence chromatin structure and accessibility, thereby regulating transcriptional activity. For example, histone acetylation generally correlates with active gene expression, while certain histone methylation marks are associated with repression [38]. The integration of genomics and epigenomics has been facilitated by large consortia such as the ENCODE Project Consortium, which aimed to identify functional elements across the human genome. ENCODE revealed that a substantial portion of the genome is biochemically active and involved in regulatory functions, challenging earlier assumptions that non-coding regions were largely non-functional [39]. Technologies such as ChIP-sequencing, bisulfite sequencing, and ATAC-sequencing have enabled genome-wide mapping of chromatin states and epigenetic landscapes.

Importantly, epigenomic modifications are dynamic and responsive to environmental influences such as diet, stress, toxins, and aging. This plasticity explains how environmental factors can shape gene expression profiles and contribute to disease susceptibility without altering the underlying DNA sequence [40]. Thus, epigenomics provides a mechanistic link between genotype and phenotype. In summary, the progression from genomics to epigenomics reflects an expanded understanding of biological complexity. While genomics deciphers the blueprint of life, epigenomics interprets how that blueprint is regulated and executed in specific cellular contexts. Together, these fields are transforming precision medicine, enabling biomarker discovery, targeted therapies, and personalized healthcare strategies.

1.3 The Data Revolution: Evolution of Sequencing Technologies

The fuel for the AI engine in genomics is data, specifically the sequence of nucleotide bases (A, C, G, T) that make up the genome. This history of DNA sequencing is a chronicle of exponential technological improvement, characterized by a transition from laborious, manual method to automated, high-throughput platforms that generate terabytes of data per run [41]. This evolution is typically categorized into three "generations" of technology.

1.3.1 First Generation: Sanger Sequencing

Developed by Frederick Sanger in 1977, the Sanger sequencing method (dideoxy chain termination) was the gold standard for three decades [41]. It relies on the use of chemically modified nucleotides (ddNTPs) that terminate DNA strand elongation. By separating these terminated fragments by size using capillary electrophoresis, the sequence can be read.

- **Characteristics**: Sanger sequencing is renowned for its high accuracy (99.99%) and relatively long read lengths (up to 1000 bp). However, it suffers from extremely low throughput and high cost per base [42]. It is fundamentally a serial process, reading one DNA fragment at a time.
- **Legacy**: Despite its limitations, Sanger sequencing was the workhorse of the Human Genome Project (HGP). This international effort, which ran from 1990 to 2003, relied on fleets of Sanger sequencers to generate the first reference genome. The project took 13 years and cost approximately $3 billion [43]. Today, Sanger sequencing is largely relegated to small-scale verification tasks where accuracy is paramount and volume is low [44].

1.3.2 Second Generation: Next-Generation Sequencing (NGS)

The term "Next-Generation Sequencing" (NGS), or Massively Parallel Sequencing, refers to the technologies that emerged in the mid-2000s and fundamentally changed the economics of genomics [41]. The dominant player in this space is Illumina.

- **Mechanism (Sequencing by Synthesis)**: Illumina's technology involves fragmenting DNA into short pieces and attaching them to a glass slide (flow cell). Through a process called "bridge amplification," single DNA molecules are amplified into clusters containing thousands of identical copies. The sequencer then floods the slide with fluorescently labeled nucleotides. As each base is incorporated, it emits a specific color signal which is captured by high-resolution cameras [41].
- **Specs and Throughput**: The defining feature of NGS is parallelism. Instead of sequencing one fragment, it sequences millions or billions simultaneously. A modern machine like the NovaSeq X can generate terabytes of data in 24 h [44]. The trade-off is read length; Illumina typically produces short reads of 150–300 bp [42].
- **Limitations**: The short-read length makes it difficult to resolve repetitive regions of the genome or complex structural variants, as short reads cannot bridge across large repeats. It is like trying to assemble a puzzle where all the blue-sky pieces look identical [42].

1.3.3 Third Generation: Long-Read Sequencing

To address the limitations of short-read sequencing, third-generation technologies emerged, capable of reading single molecules of DNA without the need for amplification (PCR), often producing reads tens of kilobases in length [45].

- **Pacific Biosciences (PacBio) HiFi**: This technology uses single molecule real-time (SMRT) sequencing. DNA polymerase is immobilized at the bottom of a microscopic well called a zero-mode waveguide (ZMW). As the polymerase incorporates phospho-linked nucleotides into a single DNA template, light pulses are emitted and detected in real time [45].
- **HiFi Reads**: A major innovation was the introduction of "HiFi" reads. By circularizing the DNA template, the polymerase can read the same molecule multiple times. While the raw error rate of a single pass is high, the consensus of multiple passes yields accuracies > 99.9% (comparable to Sanger) with read lengths of 15,000–20,000 bp [44]. This combination of length and accuracy makes HiFi the current gold standard for de novo genome assembly and variant phasing [56].

- **Oxford Nanopore Technologies (ONT)**: ONT utilizes a radically different approach. A single strand of DNA is pulled through a biological nanopore protein embedded in an electrically resistant membrane.
- **Mechanism**: As the DNA passes through the pore, it disrupts an ionic current flowing through the hole. The specific disruption pattern is characteristic of the nucleotide combination passing through (often referred to as "squiggles"). These raw electrical signals are then decoded by Recurrent Neural Networks (RNNs) into based sequences [45].
- **Specs**: ONT can produce ultra-long reads (up to millions of base pairs), theoretically limited only by the length of the DNA molecule itself. While historically less accurate than Illumina, recent chemistry updates (e.g., R10.4.1 flow cells) have pushed accuracy to $> 99\%$ [41]. A unique advantage is portability; the MinION sequencer is the size of a USB drive, allowing for sequencing in the field (e.g., during Ebola or COVID-19 outbreaks) [46].

1.4 The Big Data Challenge: Volume, Velocity, and Variety

The plummeting cost of sequencing has led to an explosion in data generation, positioning genomics as a formidable "Big Data" domain. In a seminal 2015 analysis, researchers from the University of Illinois and Cold Spring Harbor Laboratory projected that by 2025, genomics would rival or exceed the data generation of astronomy [48]. This prediction has largely materialized, transforming genomics into a "four-headed breast" that demands extreme resources across four dimensions: Acquisition, Storage, Distribution, and Analysis [48].

1.4.1 Volume: The Exabyte Scale

A single human genome sequence consists of approximately 3 billion base pairs. However, the raw data file generated by a sequencer is significantly larger. To ensure accuracy, genomes are sequenced at "30 × coverage" or higher, meaning every base is read an average of 30 times. This raw data, typically stored in FASTQ format, along with quality scores for every base, results in a file size of roughly 100 gigabytes (GB) per person.

- **2025 Projections**: Estimates suggested that by 2025, the field would need to store between 2 and 40 exabytes of data annually (where 1 exabyte = 1 billion gigabytes) [49].
- **Global Context**: To put this in perspective, in 2025, the world is generating roughly 400 million terabytes (0.4 zettabytes) of total data per day across all digital sectors

[50]. Genomics constitutes a growing and non-trivial fraction of this global datasphere. Unlike video data, which is often "write once, read rarely," genomic data is "cold" storage that must be frequently "thawed" and re-analyzed as new variant associations are discovered or new reference genomes are released.

1.4.2 Comparison with Tech Giants

The computational demands of genomics are often compared to those of major technology platforms to illustrate the scale [48]:

- **YouTube**: In 2015, users uploaded ~ 300 h of video per minute. By 2025, predictions placed this at ~ 1700 h per minute, requiring 1–2 exabytes of storage annually [49]. Genomics was projected to surpass this if sequencing adoption hit population scales (e.g., sequencing 2 billion people).
- **Twitter/X**: While Twitter generates massive amounts of data, the data itself (short text) is simpler to process than genomic reads. Genomic data is highly heterogeneous and lacks the convenient standardization of an.mp4 or.txt file. It often requires complex alignments (mapping millions of short reads to a reference genome) that are computationally expensive [51].

1.4.3 The Compute Bottleneck and the GPU Solution

The challenge has shifted from generating data to computing on it. While sequencing costs have dropped super-exponentially, the cost of computing (following Moore's Law) has dropped only exponentially. This divergence creates a widening gap: we can sequence genomes faster and cheaper than we can analyze them using traditional CPU-based pipelines [52].

- **Analysis Latency**: A standard GATK (Genome Analysis Toolkit) variant calling pipeline on a CPU might take 30 + hours to process a single whole genome.
- **GPU Shift**: This bottleneck is the primary catalyst for the adoption of GPU-accelerated AI. AI models, particularly deep learning, leverage the parallel processing architecture of Graphics Processing Units (GPUs). Technologies like NVIDIA Parabricks implement GATK pipelines on GPUS, reducing analysis time from days to minute (e.g., ~ 30 min for a whole genome) [53]. Furthermore, AI-based variant callers like DeepVariant convert sequencing data into image-like representations and use Convolutional Neural Networks (CNNs) to call variants with higher accuracy than statistical methods, fundamentally relying on GPU infrastructure [54].

1.5 The Rise of Artificial Intelligence in Genomics

Artificial Intelligence, particularly machine learning (ML), has been applied to biology for decades, but the last ten years have witnessed a transformation from "feature-based" ML to "end-to-end" deep learning (DL). This evolution mirrors the timeline of AI in computer vision and natural language processing (NLP), often lagging by a few years but achieving similarly revolutionary results when it finally arrives.

1.5.1 The Era of Classical Machine Learning (2000–2014)

In the years immediately following the completion of the Human Genome Project, researchers employed classical ML algorithms to make sense of the new data. Common tools included Support Vector Machines (SVMs), Random Forests, and Hidden Markov Models (HMMs) [55].

- **Applications**: These models were used for gene prediction, classifying cancer subtypes form microarray gene expression data, and identifying splice sites [55].
- **The Limitation (Feature Engineering)**: The defining characteristic of this era was the need for extensive feature engineering. A biological expert had to manually define what features the model should look for. For example, to predict a gene start site, a human would tell the model to "look for a TATAAA sequence" or "calculate the GC content." The models could not learn raw sequence patterns on their own; they were limited by the biologist's prior knowledge [56].

1.5.2 The Deep Learning Revolution (2015–Present)

The "AlexNet moment" for genomics, a reference to the 2012 breakthrough where a deep neural network crushed the competition in image recognition arrived in genomics around 2015. This paradigm shift was marked by the realization that a DNA sequence (a linear string of A, C, G, T) is mathematically analogous to a sentence in NLP (a sequence of words) or a row of pixels in an image (a 1D vector). Deep neural networks (DNNs) could learn regulatory grammar directly from the raw sequence without manual feature extraction [19].

A timeline of key publications that defined the field. Table 1.1 shows seminal milestones in AI for genomics.

The convergence of three factors drives this "AI Summer" in genomics:

- **Data Availability**The existence of massive biobanks (UK Biobank, All of Us) and public databases (PDB, UniProt) provides the necessary training data [61].

Table 1.1 Seminal milestones in AI for genomics

Year	Model/paper	Innovation	Impact
2015	DeepBind [57]	First major application of Convolutional Neural Networks (CNNs) to DNA	Predicted DNA-binding specificities of proteins directly from raw sequence, outperforming all previous feature-based methods
2015	DeepSEA [58]	Deep learning for non-coding variant effect prediction	Enabled scoring of mutations in "junk DNA" to see if they disrupted regulatory elements, effectively predicting chromatin effects from sequence alone
2017	DeepVariant [59]	Computer vision for variant calling	Google treated sequencing read pileups as "images" (RGB channels representing bases/quality) and used CNNs to identify mutations, achieving superhuman accuracy
2018	AlphaFold 1 [60]	Entry into CASP13 competition	Demonstrated that AI could compete with physics-based methods for predicting protein 3D structure, placing them first in the competition
2018	PrimateAI [56]	Trained on non-human primate genomes	Used evolutionary conservation (natural selection) as a training signal to predict pathogenicity of human missense mutations
2020	AlphaFold 2 [60]	The breakthrough. Solved the "Protein Folding Problem"	Achieved atomic-level accuracy, predicting structures for nearly all cataloged proteins. Cited over 28,000 times [61]
2021	Enformer [35]	Transformer architecture for DNA	Used attention mechanisms to model long-range interactions (up to 100 kb), connecting enhancers to promoters effectively
2023	PrimateAI-3D [62]	Integration of 3D structure and evolution	Improved identification of disease-causing variants by combining sequence conservation with AlphaFold-predicted structures
2024	AlphaFold 3 [59]	Generative biology/ interaction modeling	Expanded beyond proteins to predict interactions with DNA, RNA, and ligands (drugs), facilitating direct drug discovery applications

- **Hardware**: The availability of GPUs capable of parallel matrix operations required for training deep networks has removed the computational ceiling [53].
- **Algorithmic Maturity**: The adaptation of architectures like Transformers (originally from Google's BERT/GPT for language) to biological sequences has unlocked the ability to model long-range dependencies in the genome [63].

1.6 Neural Architectures in Biology: From CNNs to Transformers

Just as different tools are required for carpentry and plumbing, different AI architectures are suited for specific genomic tasks. The field has evolved from simple feed-forward networks to complex, attention-based models that mirror the complexity of the biological systems they model.

1.6.1 Convolutional Neural Networks (CNNs)

Originally designed for image recognition, CNNs are exceptionally well-suited for analyzing DNA sequences, particularly for motif discovery.

- **Logic**: In computer vision, a CNN scans an image for edges and shapes regardless of where they appear (translation invariance). In DNA, a transcription factor binding site (motif) like "GATAA" might appear anywhere along the gene's promoter region. A CNN uses 1D convolutional "filters" to scan the sequence. When a filter matching "GATAA" slides over that sequence, it activates, signaling the presence of the motif [64].
- **Applications**: DeepBind and DeepSEA are classic examples. DeepSEA uses CNNs to scan 1000 bp sequences of the genome and predict whether that region is accessible (open chromatin) or bound by specific transcription factors [58].
- **Strengths**: Excellent at detecting local patterns (motifs) and short-range interactions. They are computationally efficient compared to Transformers [65].

1.6.2 Recurrent Neural Networks (RNNs) and LSTMs

RNNs are designed for sequential data where the order matters and the context of previous elements influences the current one (like time-series or text).

- **Logic**: They process DNA base-by-base, retaining a "memory" (hidden state) of previous bases. This is particularly useful for tasks like nanopore base calling, where the raw electrical signal stream must be decoded into a sequence of A, C, G, T. The signal at time t depends on the nucleotides that passed through the pore at $t - 1$ [37].
- **Limitations**: Traditional RNNs struggle with very long sequences due to the "vanishing gradient problem," where the model forgets information from the beginning of a long sequence. This makes them less effective for modeling interactions between distant parts of the genome (e.g., an enhancer 50,000 bases away from its promoter).

1.6.3 Transformers and Attention Mechanisms

The introduction of the Transformer architecture (originally for language translation, e.g., "The cat sat on the mat") revolutionized genomic AI by solving the long-range dependency problem.

- **Logic**: Instead of processing data sequentially (RNN) or locally (CNN), Transformers use a Self-Attention Mechanism. This allows the model to "attend" to all parts of the sequence simultaneously. It can calculate the relevance (weight) of a base at position 1 relative to a base at position 100,000. In biological terms, it can "see" that an enhancer at position X is regulating a gene at position Y, even if they are separated by vast stretches of non-coding DNA [65].
- **Applications (Enformer and Nucleotide Transformer)**: Enformer uses this architecture to predict gene expression by integrating information from up to 100 kb of context, significantly outperforming CNN-only models in predicting the impact of non-coding variants [35].
- **Impact**: This capability is critical for interpreting the "dark matter" of the genome. It moves the field closer to a "genomic language model" that understands the syntax of gene regulation [53].

1.6.4 Graph Neural Networks (GNNs)

GNNs operate on data structured as graphs (nodes and edges) rather than linear sequences or grids.

- **Logic**: Biology is often defined by structures and networks. A protein is a graph where amino acids are nodes and their physical proximity in 3D space forms the edges. A metabolic pathway is a graph of molecules and enzymes. GNNs can learn from these non-Euclidean structures [66].
- **Applications (Protein Structure Prediction and Drug Discovery)**: AlphaFold 2 utilizes a variation of attention that has graph-like properties (the Evoformer module) to reason about the 3D spatial relationship of residues, effectively treating the protein folding problem as a graph inference task. GNNs are also used to predict the properties of small molecule drugs and their interaction with protein targets [67].

1.7 Ethical Frontiers and Future Horizons

As AI and genomics converge, they generate ethical and technical challenges that are as complex as science itself. The power to read and write the code of life carries profound responsibilities.

1.7.1 The Bias Bottleneck

A major failure mode of current genomic AI is bias in training data. Most genomic studies (GWAS) have been conducted on populations of European ancestry; estimates suggest over 80% of GWAS participants are of European descent, despite making up a fraction of the global population [68].

- **Consequence**: AI models trained on this biased data inherit these limitations. A polygenic risk score (PRS) trained on European data to predict breast cancer or heart disease risk performs significantly worse when applied to African, Asian, or Hispanic populations. This effectively creates a two-tiered healthcare system where AI diagnostics work for some but fail for others, exacerbating existing health disparities [69].
- **Solution**: Initiatives like the All of Us research program and specific AI techniques (like Transfer Learning) are attempting to mitigate this. Transfer learning allows a model to learn general features from a large, biased dataset and then be "fine-tuned" on a smaller, diverse dataset, but "de-biasing" genomic AI remains a primary challenge for the next decade [70].

1.7.2 Privacy and the Genome

Genomic data is uniquely identifying; you cannot "anonymize" a genome in the traditional sense because the sequence itself is the ultimate identifier.

- **Privacy Attacks**: Research has demonstrated the vulnerability of genomic models to privacy attacks. Membership Inference Attacks allow an adversary to query a machine learning model (e.g., a cancer prediction model) and statistically determine if a specific individual's data was used to train it [71]. Model Inversion Attacks can potentially reconstruct sensitive genetic markers of the training participants by analyzing the model's outputs [71].
- **Implication**: This creates a tension between the need for open data to train massive AI models (which require millions of genomes) and the need to protect patient privacy.

Techniques like Federated Learning, where models are trained locally on decentralized data servers and only the model updates (not the data) are shared, are being explored as potential solutions [72].

1.7.3 The Future: Generative Biology and Synthetic Design

Looking toward 2026 and beyond, the field is moving from "reading" biology (predicting properties of existing sequences) to "writing" it (designing new biological entities).

- **Generative AI**: Just as tools like MidJournary or Stable Diffusion generate images from text prompts, new generative AI models in biology (e.g., ProteinMPNN, RFdiffusion) are generating novel protein sequences that do not exist in nature. These "hallucinated" proteins can be designed to perform specific functions, such as binding to a specific cancer marker or catalyzing a specific chemical reaction [19].
- **Success Stories**: We are already seeing the first waves of AI-designed drugs entering and succeeding in clinical trials. In 2024/25, Insilico Medicine's AI-designed drug for Idiopathic Pulmonary Fibrosis (IPF), Rentosertib, successfully passed Phase IIa trails. This marked a historic validation of end-to-end AI drug discovery, where AI identified the target, designed the molecule, and predicted the clinical trial outcome [73].

The integration of Genomics and AI is not merely an incremental improvement in biological research; it is a rewriting of the rules of discovery. We are moving from a hypothesis-driven science, where we ask specific questions based on limited knowledge, to a data-driven science, where AI illuminates the questions we didn't even know to ask.

1.8 Summary

The chapter provides an in-depth introduction to the intersection of genomics and artificial intelligence (AI), highlighting the shift from genetics to genomics and the transformative role of AI in modern biology. Genetics focuses on individual genes and their inheritance, while genomics studies the entire genome, including complex interactions between genes and environmental factors. This shift to multifactorial analysis has made AI indispensable for analyzing vast genomic datasets, identifying patterns, and understanding polygenic diseases. The chapter explains the central dogma of molecular biology (DNA $\rightarrow$ RNA $\rightarrow$ Protein) and delves into the complexities of gene architecture, regulatory elements like promoters and enhancers, and the challenges of predicting gene expression. It also explores the evolution of DNA sequencing technologies, from Sanger sequencing to next-generation and long-read sequencing, which have drastically increased data generation. The chapter emphasizes the role of AI in overcoming computational bottlenecks, with advancements in deep learning, CNNs, RNNs, transformers, and Graph Neural Networks revolutionizing genomic research. Ethical challenges, such as bias in training data and privacy concerns, are discussed, alongside the future potential of generative biology and AI-driven drug discovery. Overall, the chapter underscores how AI is reshaping genomics into a data-driven science, enabling breakthroughs in understanding and manipulating the code of life.

References

1. Watson, J. D., & Crick, F. H. C. (1953). Molecular structure of nucleic acids: A structure for deoxyribose nucleic acid. *Nature, 171*(4356), 737–738. https://doi.org/10.1038/171737a0
2. Mullis, K., & Faloona, F. (1987). Specific synthesis of DNA in vitro via a polymerase-catalyzed chain reaction. *Methods in Enzymology, 155*, 335–350. https://doi.org/10.1016/0076-6879(87)55023-6
3. Sanger, F., Nicklen, S., & Coulson, A. R. (1977). DNA sequencing with chain-terminating inhibitors. *Proceedings of the National Academy of Sciences, 74*(12), 5463–5467. https://doi.org/10.1073/pnas.74.12.5463
4. Lander, E. S., Linton, L. M., Birren, B., Nusbaum, C., Zody, M. C., Baldwin, J., Devon, K., Dewar, K., Doyle, M., FitzHugh, W., Funke, R., Gage, D., Harris, K., Heaford, A., Howland, J., Kann, L., Lehoczky, J., LeVine, R., McEwan, PP., ... (2001). International Human Genome Sequencing Consortium. Initial sequencing and analysis of the human genome. *Nature, 409*(6822), 860–921. https://doi.org/10.1038/35057062
5. Venter, J. C., Adams, M. D., Myers, E. W., Li, P. W., Mural, R. J., Sutton, G. G., Smith, H. O., Yandell, M., Evans, C. A., Holt, R. A., Gocayne, J. D., Amanatides, P., Ballew, R. M., Huson, D. H., Wortman, J. R., Zhang, Q., Kodira, C. D., Zheng, X. H., Chen, L., ... & Zhu, X. (2001). The sequence of the human genome. *Science, 291*(5507), 1304–1351. https://doi.org/10.1126/science.1058040
6. Metzker, M. L. (2010). Sequencing technologies—The next generation. *Nature Reviews Genetics, 11*(1), 31–46. https://doi.org/10.1038/nrg2626
7. Patel, A., Gupta, V., & Shah, S. (2021). Metabolomic profiling of an Indian oral squamous cell carcinoma subset. *Analytical Chemistry Letters, 11*(4), 596–606. https://doi.org/10.1080/22297928.2021.1936168
8. Patel, A., Kikani, A., Makadiya, G., & Shah, S. (2024). Transcriptomic profiling of OSCC patients in an Indian subset. *Asian Pacific Journal of Cancer Prevention, 25*(1), 233–239. https://doi.org/10.31557/APJCP.2024.25.1.233
9. Patel, A., Mahapatra, S., Bishoyi, A. K., Sharma, A., Makwana, A., Swarnkar, T., Gupta, A., Sahoo, P. K., & Shah, S. (2025). Harnessing machine learning technique to authenticate differentially expressed genes in oral squamous cell carcinoma. *Oral Surgery, Oral Medicine, Oral Pathology and Oral Radiology, 139*(2), 211–219. https://doi.org/10.1016/j.oooo.2024.10.075
10. Shah, S. (2018). Single nucleotide polymorphism rs17849071 G/T in the PIK3CA gene is inversely associated with oral cancer. *Oral Cancer, 2*, Article 2. https://doi.org/10.1007/s41548-018-0012-1
11. Shah, S., Jajal, D., Mishra, G., & Kalia, K. (2016). Genetic profile of PTEN gene in Indian oral squamous cell carcinoma primary tumors. *Journal of Oral Pathology & Medicine, 46*(2), 106–111. https://doi.org/10.1111/jop.12468

12. Shah, S., Padh, H., & Kalia, K. (2015). Genetic alterations of the PIK3CA oncogene in human oral squamous cell carcinoma in an Indian population. *Oral Surgery, Oral Medicine, Oral Pathology and Oral Radiology, 120*(5), 628–635. https://doi.org/10.1016/j.oooo.2015.08.003
13. Patel, A., Bavarva, A., Swarnkar, T., Gupta, V., Trivedi, Y., & Shah, S. (2025). Ensemble Learning Approaches in HPV Associated Oral Squamous Cell Carcinoma. 2025 IEEE 13th Region 10 Humanitarian Technology Conference (R10-HTC), 123–127. https://doi.org/10.1109/R10-HTC63995.2025.11394628
14. Williams, G. A., Liede, S., Fahy, N., Aittomaki, K., Perola, M., Helander, T., ... & Sagan, A. (2021). *Regulating the unknown: A guide to regulating genomics for health policy-makers*. https://www.ncbi.nlm.nih.gov/books/NBK569502/
15. Genetics vs. Genomics Fact Sheet. https://www.genome.gov/about-genomics/fact-sheets/Genetics-vs-Genomics. Last Access: January 2026.
16. Genetics vs. Genomics—The Jackson Laboratory. https://www.jax.org/personalized-medicine/precision-medicine-and-you/genetics-vs-genomics. Last Access: January 2026.
17. Your Guide to Genetics and Genomics—Stanford Online. https://online.stanford.edu/your-guide-genetics-and-genomics. Last Access: January 2026.
18. Genetics vs. Genomics: What's the Difference? (2017). https://blog.dana-farber.org/insight/2016/10/genetics-vs-genomics-whats-the-difference/. Last Access: January 2026.
19. Lee, D. H., Park, E. G., Lee, Y. J., Jeong, H. S., Roh, H. Y., Jeong, G. R., … & Kim, H. S. (2025). Advances in Artificial Intelligence (AI) models and generative algorithms represent a new paradigm for genomics research. *International Journal of Molecular Sciences, 26*(22), 10925.
20. DNA Replication, Transcription and Translation. (2023). https://adntro.com/en/blog/learn-genetics/replication_transcription_translation/. Last Access: January 2026.
21. Crick, F. H. C. (1958). On protein synthesis. In *Symposia of the Society for Experimental Biology; Number XII: The Biological Replication of Macromolecules*.
22. How is DNA turned into protein? The Central Dogma of Molecular Biology. https://www.yourgenome.org/theme/how-is-dna-turned-into-protein-the-central-dogma-of-molecular-biology/. Last Access: January 2026.
23. Martinez, M. (2022). *The central dogma of molecular biology*. https://www.zeclinics.com/blog/relationship-between-dna-rna-and-proteins/. Last Access: January 2026.
24. Alberts, B., Johnson, A., Lewis, J., Morgan, D., Raff, M., Roberts, K., & Walter, P. (2015). Molecular biology of the cell (6th ed.). Garland Science.
25. Berg, J. M., Tymoczko, J. L., Gatto, G. J., & Stryer, L. (2019). Biochemistry (9th ed.). W. H. Freeman.
26. Lodish, H., Berk, A., Kaiser, C. A., Krieger, M., Scott, M. P., Bretscher, A., Ploegh, H., & Matsudaira, P. (2016). Molecular cell biology (8th ed.). W. H. Freeman.
27. Differential Gene Transcription—Developmental Biology. (2000). https://www.ncbi.nlm.nih.gov/books/NBK10023/. Last Access: January 2026.
28. Kornblihtt, A. R., Schor, I. E., Alló, M., Dujardin, G., Petrillo, E., & Muñoz, M. J. (2013). Alternative splicing: a pivotal step between eukaryotic transcription and translation. *Nature reviews Molecular cell biology, 14*(3), 153–165.
29. Bulger, M., & Groudine, M. (2011). Functional and mechanistic diversity of distal transcription enhancers. *Cell, 144*(3), 327–339.
30. Maston, G. A., Evans, S. K., & Green, M. R. (2006). Transcriptional regulatory elements in the human genome. *Annual Review of Genomics and Human Genetics, 7*(1), 29–59.
31. Bartel, D. P. (2004). MicroRNAs: genomics, biogenesis, mechanism, and function. *Cell, 116*(2), 281–297.

32. Anna, A., & Monika, G. (2018). Splicing mutations in human genetic disorders: examples, detection, and confirmation. *Journal of Applied Genetics, 59*(3), 253–268.
33. Transcription, Translation and Replication—Nucleic Acids Book. https://atdbio.com/nucleic-acids-book/Transcription-Translation-and-Replication. Last Access: January 2026.
34. Enhancer (genetics)—Wikipedia. https://en.wikipedia.org/wiki/Enhancer_(genetics). Last Access: January 2026.
35. Implementation of Enformer, Deepmind's attention network for predicting gene expression in pytorch. GitHub. https://github.com/lucidrains/enformer-pytorch. Last Access: January 2026.
36. International Human Genome Sequencing Consortium. (2004). Finishing the euchromatic sequence of the human genome. *Nature, 431*(7011), 931–945.
37. Goodwin, S., McPherson, J. D., & McCombie, W. R. (2016). Coming of age: ten years of next-generation sequencing technologies. *Nature Reviews Genetics, 17*(6), 333–351.
38. Allis, C. D., & Jenuwein, T. (2016). The molecular hallmarks of epigenetic control. *Nature Reviews Genetics, 17*(8), 487–500.
39. ENCODE Project Consortium. (2012). An integrated encyclopedia of DNA elements in the human genome. *Nature, 489*(7414), 57–74.
40. Feil, R., & Fraga, M. F. (2012). Epigenetics and the environment: emerging patterns and implications. *Nature Reviews Genetics, 13*(2), 97–109.
41. Mobley, I. (2021). *A Brief History of Next Generation Sequencing (NGS).* https://frontlinegenomics.com/a-brief-history-of-next-generation-sequencing-ngs/. Last Access: January 2026.
42. Meszaros, E. (2023). *Sanger Sequencing vs NGS.* https://www.integra-biosciences.com/united-states/en/blog/article/sanger-sequencing-vs-ngs. Last Access: January 2026.
43. A Journey Through the History of DNA Sequencing. https://the-dna-universe.com/2020/11/02/a-journey-through-the-history-of-dna-sequencing/. Last Access: January 2026.
44. Genomics Gets Faster, Cheaper, and More Accurate. https://sangerinstitute.blog/2024/02/29/genomics-gets-faster-cheaper-and-more-accurate/. Last Access: January 2026.
45. Cox, R., & Timilsina, S. (2023). *Ths history of DNA sequencing.* https://www.criver.com/eureka/history-dna-sequencing. Last Access: January 2026.
46. Sequencing 101: Comparing Long-Read Sequencing Technologies. (2024). https://www.pacb.com/blog/sequencing-101-comparing-long-read-sequencing-technologies/. Last Access: January 2026.
47. Wetterstrand, K. (2023). *DNA Sequencing Costs: Data.* https://www.genome.gov/about-genomics/fact-sheets/DNA-Sequencing-Costs-Data. Last Access: January 2026.
48. Stephens, Z. D., Lee, S. Y., Faghri, F., Campbell, R. H., Zhai, C., Efron, M. J., … & Robinson, G. E. (2015). Big data: Astronomical or genomical? *PLoS Biology, 13*(7), e1002195.
49. Taylor, N. (2015). *Genomics vs. YouTube: Which will have the greater computing requirements in 2025?* https://www.fiercebiotech.com/data-management/genomics-vs-youtube-which-will-have-greater-computing-requirements-2025. Last Access: January 2026.
50. Duarte, F. (2025). *Amount of data created daily.* https://explodingtopics.com/blog/data-generated-per-day. Last Access: January 2026.
51. Ahlberg, L. (2015). *Genomics to surpass the biggest data producers, experts warn.* https://news.illinois.edu/genomics-to-surpass-the-biggest-data-producers-experts-warn/. Last Access: January 2026.
52. Muir, P., Li, S., Lou, S., Wang, D., Spakowicz, D. J., Salichos, L., … & Gerstein, M. (2016). The real cost of sequencing: Scaling computation to keep pace with data generation. *Genome Biology, 17*(1), 53.
53. Every, L. (2025). *The power of AI in genomics: A collaborative effort to build the genome ocean LLM.* https://it.lbl.gov/the-power-of-ai-in-genomics-a-collaborative-effort-to-build-the-genome ocean-llm/. Last Access: January 2026.

54. DeepVariant. https://github.com/google/deepvariant. Last Access: January 2026.
55. Caudai, C., Galizia, A., Geraci, F., Le Pera, L., Morea, V., Salerno, E., … & Colombo, T. (2021). AI applications in functional genomics. *Computational and Structural Biotechnology Journal, 19*, 5762–5790.
56. Sundaram, L., Gao, H., Padigepati, S. R., McRae, J. F., Li, Y., Kosmicki, J. A., … & Farh, K. K. H. (2018). Predicting the clinical impact of human mutation with deep neural networks. *Nature Genetics, 50*(8), 1161–1170.
57. Lan, A. Y., & Corces, M. R. (2022). Deep learning approaches for noncoding variant prioritization in neurodegenerative diseases. *Frontiers in Aging Neuroscience, 14*, 1027224.
58. Li, Z., Gao, E., Zhou, J., Han, W., Xu, X., & Gao, X. (2023). Applications of deep learning in understanding gene regulation. *Cell Reports Methods, 3*(1).
59. AlphaFold—Google DeepMind. https://deepmind.google/science/alphafold/. Last Access: January 2026.
60. AlphaFold: Using AI for Scientific Discovery—Google DeepMind. https://deepmind.google/blog/alphafold-using-ai-for-scientific-discovery-2020/. Last Access: January 2026.
61. Oana Stroe. *Case study: AlphaFold uses open data and AI to discover the 3D protein universe.* https://www.embl.org/news/science/alphafold-using-open-data-and-ai-to-discover-the-3d-protein-universe/. Last Access: January 2026.
62. Illumina Unveils AI Software to Predict Disease-Causing Genetic Mutations in Patients (2023). https://www.illumina.com/company/news-center/press-releases/press-release-details.html?newsid=b067ea3b-be0e-4318-a910-6bd659e0d232. Last Access: January 2026.
63. Shen, X., Jiang, C., Wen, Y., Li, C., & Lu, Q. (2022). A brief review on deep learning applications in genomic studies. *Frontiers in Systems Biology, 2*, 877717.
64. Wang, X., Li, F., Zhang, Y., Imoto, S., Shen, H. H., Li, S., ... & Song, J. (2024). Deep learning approaches for non-coding genetic variant effect prediction: current progress and future prospects. *Briefings in Bioinformatics, 25*(5), bbae446.
65. Manzo, G., Borkowski, K., & Ovcharenko, I. (2025). Comparative analysis of deep learning models for predicting causative regulatory variants. *Genes, 16*(10), 1223.
66. Van Hilten, A., Katz, S., Saccenti, E., Niessen, W. J., & Roshchupkin, G. V. (2024). Designing interpretable deep learning applications for functional genomics: a quantitative analysis. *Briefings in Bioinformatics, 25*(5).
67. Calvino, G., Farro, J., Zampatti, S., Peconi, C., Megalizzi, D., Trastulli, G., … & Giardina, E. (2025). From genomics to AI: Revolutionizing precision medicine in oncology. *Applied Sciences, 15*(12), 6578.
68. Vilhekar, R. S., & Rawekar, A. (2024). Artificial intelligence in genetics. *Cureus, 16*(1).
69. Norori, N., Hu, Q., Aellen, F. M., Faraci, F. D., & Tzovara, A. (2021). Addressing bias in big data and AI for health care: A call for open science. *Patterns, 2*(10).
70. The Role of AI in Genomics: Opportunities and Challenges. https://publicpolicyprojects.com/the-role-of-ai-in-genomics-opportunities-and-challenges/. Last Access: January 2026.
71. Chen, J., Wang, W. H., & Shi, X. (2020). Differential privacy protection against membership inference attack on machine learning for genomic data. In *Biocomputing 2021: Proceedings of the Pacific Symposium* (pp. 26–37).
72. Johnson, K. B., Wei, W. Q., Weeraratne, D., Frisse, M. E., Misulis, K., Rhee, K., … & Snowdon, J. L. (2021). Precision medicine, AI, and the future of personalized health care. *Clinical and Translational Science, 14*(1), 86–93.
73. Accelerating Drug Development with AI in the U.S. Pharmaceutical Industry. https://intuitionlabs.ai/articles/accelerating-drug-development-ai-pharma. Last Access: January 2026.

Genomic Data and Computational Foundations

2

The transformation of biology into a data-driven discipline has been precipitated by a singular, overwhelming force: the exponential decrease in the cost of nucleotide sequencing. As the "read" becomes the fundamental unit of biological observation, the focus of scientific inquiry has expanded from the qualitative study of individual genes to the quantitative interrogation of entire genomes, transcriptomes, and epigenomes. This shift necessitates a rigorous foundational understanding of the data itself, its physical origins, its digital encodings, and the mathematical transformations required to extract meaning from noise.

This chapter establishes the technical bedrock for modern genomic analysis. It moves beyond the biological intuition of the wet lab to the precise specifications of computational science. We will dissect the hierarchy of genomic data types, analyzing the distinct error profiles and signal properties of short-read versus long-read technologies. We will rigorously define the file formats that serve as the lingua franca of bioinformatics, clarifying the often-misunderstood coordinate systems and bitwise flags that underpin data interoperability. Furthermore, we will explore the mathematical logic of normalization, the architecture of high-performance computing required to process petabyte-scale datasets, and the emerging paradigm of pangenomes that challenge our linear conception of reference genomes. This foundation is essential for constructing the scalable, reproducible pipelines discussed in the final sections, enabling researchers to navigate the transition from raw signal to biological insight with precision and confidence.

K. Shaikh et al., *Artificial Intelligence in Genomics*, Synthesis Lectures on Engineering, Science, and Technology, https://doi.org/10.1007/978-3-032-25486-3_2

2.1 Types of Genomic Data

The term "genomic data" acts as a convenient abstraction for a complex array of measurement modalities. The digital representation of biological molecules varies profoundly depending on the underlying chemistry of the sequencing platform and the specific molecular layer being assayed. To design effective analytical strategies, one must first appreciate the distinct statistical properties, biases, and information content inherent to each data type [1].

2.1.1 DNA Sequencing: The Dichotomy of Length and Accuracy

The primary cleavage in genomic data exists between Second-Generation (Short-Read) and Third-Generation (Long-Read) sequencing. This distinction is not merely technical but dictates the class of biological questions that can be answered [1].

Short-Read Sequencing (Next-Generation Sequencing (NGS))—Illumina sequencing, the dominant paradigm for the past decade, relies on a sequencing by synthesis chemistry. DNA fragments are amplified into clusters on a flow cell, and fluorescently labeled nucleotides are incorporated cycle by cycle. This process generates "reads" typically ranging from 75 to 300 base pairs (bp) in length. The defining characteristics of short-read data are its high per based accuracy, typically exceeding 99.9%. This low error rate makes short reads the gold standard for identifying small nucleotide variants (SNVs) and small insertions/deletions (Indels) in non-repetitive regions. However, the short-read length poses a fundamental limitation: these readings cannot span repetitive elements longer than the reading itself. Consequently, genomic regions rich in segmental duplications, centromeres, or complex structural variants remain invisible or poorly resolved in short-read assemblies [1]. Within the short-read ecosystem, experimental design dictates the scope of inquiry:

- **Whole Genome Sequencing (WGS)**: This approach attempts to sequence the entire distinct sequence of the genome. By covering both coding and non-coding regions, WGS allows for the detection of regulatory variants and structural variants (SVs) defined by read-depth alterations or discordant read pairs. It provides the most comprehensive view of the genome but imposes the highest computational and storage burdens.
- **Whole Exome Sequencing (WES)**: By using hybridization probes to capture only the protein-coding regions (exons) approximately 1–2% of the human genome. WES offers a cost-effective alternative for identifying Mendelian disease variants. However, this cost savings comes at the price of context; WES is blind to deep intronic mutations, promoter variants, and most structural rearrangements, significantly limiting its utility in complex trait association studies [2].

Long-Read Sequencing (Third Generation Sequencing) to overcome the assembly fragmentation causes by short reads, long-read technologies like Pacific Biosciences (PacBio) and Oxford Nanopore Technologies (ONT) were developed. These platforms generate reads ranging from 10 kilobases (kb) to over 1 megabase (Mb) [1].

- **Pacific Biosciences (PacBio) HiFi**: PacBio utilizes single molecule real-time (SMRT) sequencing, where a polymerase anchored at the bottom of a zero-mode waveguide (ZMW) incorporates nucleotides. The breakthrough "HiFi" (high fidelity) mode involves circularizing the DNA template and reading it repeatedly (circular consensus sequencing). By averaging these multiple passes, random errors are corrected, yielding reads that are both long (15–20 kb) and highly accurate (> 99.9%). This accuracy allows for precise phasing of haplotypes and the resolution of highly homologous regions [1].
- **Oxford Nanopore Technologies (ONT)**: This platform measures disturbances in ionic current as a single DNA molecule is ratcheted through a protein nanopore embedded in a membrane. The signal is computationally decoded (base-called) into a sequence. While historically possessing higher error rates, particularly in homopolymer regions, ONT offers unique advantages. It supports ultra-long reads (N50 > 100 kb) essential for scaffolding Telomere-to-Telomere (T2T) assemblies. Crucially, the raw electrical signal contains information about based modifications. This allows for the direct detection of epigenetic marks, such as 5-methylcytosine (5 mC), without the need for bisulfite conversion, unifying genomic and epigenomic profiling in a single assay [2].

2.1.2 Transcriptomics: Quantifying Cellular Dynamics

While the genome represents the static potential of an organism, the transcriptome reflects its dynamic realization. Transcriptomics, usually assayed via RNA-seq, quantifies the presence and abundance of RNA molecules, providing a snapshot of cellular state tailored to specific tissues, developmental stages, or environmental stimuli [1].

Bulk RNA-sequencing standard "bulk" RNA-seq measures the average expression levels across a population of cells. The preparation protocol significantly influences the data's interpretation:

- **Poly-A Selection**: This method enriches for messenger RNAs (mRNAs) containing polyadenylated tails, effectively filtering out the abundant ribosomal RNA (rRNA). It focuses the analysis on the protein-coding transcriptome but omits non-coding regulatory RNAs.

- **rRNA Depletion (Ribozero)**: By removing ribosomal RNA, this method retains both mRNAs and diverse non-coding RNAs (Inc RANs, snRNAs) as well as nascent transcripts. This provides a more complete view of transcriptional activity but requires deeper sequencing to achieve comparable sensitivity for coding genes [2].

Single-cell RNA-sequencing (scRNA-seq) obscures cellular heterogeneity, averaging out rare cell populations. Single-cell RNA-seq resolves this by partitioning individual cells (via microfluidic droplets or plates) and tagging their transcripts with unique cellular barcodes. This resulting data is a high dimensional, extremely sparse matrix (genes x cells), where "dropout" events failure to capture a transcript are common. This necessitates specialized computational frameworks involving zero-inflated models and dimensionality reduction techniques to identify cell types and trajectories [3].

2.1.3 Epigenomics: The Layer of Regulation

Epigenomics data captures the chemical and physical modifications of DNA and chromatin that regulate gene expression without altering the sequence itself. These essays are critical for understanding the mechanism of cell differentiation and disease states where the genome remains wild type, but regulation goes awry [1].

Methylation of cytosine at CpG dinucleotides is a key repressive mark. Whole Genome Bisulfite Sequencing (WGBS) is the standard assay. Sodium bisulfite treatment converts unmethylated cytosines to uracil (sequenced as thymine) while protecting methylated cytosines. Aligners must handle a "three-letter" genome, mapping reads to parallel reference versions (C-to-T converted and G-to-A converted). This increases the search space and alignment ambiguity [2].

The assay for transposase-accessible chromatin using sequencing utilizes a hyperactive Tn5 transposase to insert sequencing adapters specifically into open, accessible regions of chromatin. The data manifests as "peaks" pileups of reads indicating active regulatory elements like promoters and enhancers. Unlike ChIP-seq, which queries a specific protein, ATAC-seq provides a global view of the regulatory landscape [3].

Chromatin Immunoprecipitation sequencing uses antibodies to isolate DNA fragments bound by specific proteins (transcription factors) or modified histones (e.g., H3K4me3). It provides direct evidence of regulatory binding but relies heavily on antibody specificity and signal-to-noise ratios [4]. Chromosomes are not linear threads but folded globules. Hi-C captures physical contact between genomic loci, generating interaction matrices. These maps reveal topologically associating domains (TADs) and loops, explaining how enhancers can regulate promoters located megabases away in the linear sequence [5].

2.1.4 Metagenomics and Multi-Omics

The scope of genomic data extends beyond the individual organism. Metagenomics involves sequencing DNA directly from environmental samples (e.g., gut microbiome, soil), bypassing the need for culturing. This data is characterized by high diversity and uneven coverage, requiring "binning" algorithms to reconstruct genomes from complex mixtures [2].

Finally, the frontier of multi-omics involves the integration of these distinct layers. By assaying the genome, epigenome, and transcriptome simultaneously from the same biological samples (or even the same cells), researchers can construct causal models of regulation linking a genetic variant (genomics) to an enhancer state (epigenomics) and ultimately to gene output (transcriptomics) [1].

2.2 Sequencing Data Formats

Bioinformatics is a discipline defined by its file formats. The efficiency, accuracy, and interoperability of genomic analysis depend entirely on the rigorous specification of these data containers. A computational biologist must understand not just how to use these files, but how they are structured at the bit and byte level.

2.2.1 Raw Sequence Data: The FASTQ Standard

The FASTQ format is the universal starting point for high-throughput sequencing analysis. It is a text-based format that extends the simple sequence representation of FASTA by adding a parallel stream of quality data [6].

- **Structure of a FASTQ Record**: A single read is represented by a quartet of lines:
  - **Header**: Initiated by "@" symbol, containing the sequence identifier (e.g., SRR001666.1). This line typically encodes the instrument ID, run ID, flow cell lane, tile, and coordinates, providing a unique address for every cluster.
  - **Sequence**: The raw base calls (A, C, G, T, N).
  - **Separator**: A + sign, occasionally followed by the identifier again for redundancy.
  - **Quality String**: A sequence of ASCII characters matching the length of the DNA sequence, encoding the confidence of each base call [7].
- **Phred Quality Scores**: The quality of a base call is expressed mathematically as a Phred score (Q) which is logarithmically related to the probability of error (P):

$$Q = -10 \log_{10} P$$

This logarithmic scale means that a score of $\boldsymbol{Q_{10}}$ implies a 1 in 10 chance of error (90% accuracy), $\boldsymbol{Q_{20}}$ implies a 1 in 100 (99% accuracy), and $\boldsymbol{Q_{30}}$ implies a 1 in 1,000 (99.9% accuracy). In the context of billion-read datasets, $\boldsymbol{Q_{30}}$ **is** generally considered the threshold for high-quality data [7].

- **Encoding Schemas: Phred + 33 versus Phred + 64**

To represent the integer $\boldsymbol{Q}$ (ranging from 0 to ~ 40 or 93) as a single printable character, an ASCII offset is applied.

- **Phred + 33 (Sanger/Illumina 1.8+)**: This standard for all modern data. The score $\boldsymbol{Q}$ is added to 33. Thus, $\boldsymbol{Q = 0}$ is ASCII 33 (!), and $\boldsymbol{Q = 40}$ is ASCII 73 (I).
- **Phred + 64 (Illumina 1.3–1.7)**: A legacy format where the offset was 64. $\boldsymbol{Q = 0}$ is ASCII 64 (@), and $\boldsymbol{Q = 40}$ is ASCII 104 (h). Distinguishing these formats is critical. Misinterpreting a Phred + 64 file as Phred + 33 will result in heavily inflated quality scores, causing bad data to pass filtration filters. Conversely, the reverse error will cause good data to be discarded. Heuristics usually check for characters unique to one range (e.g., characters 35–58 imply Phred + 33; characters > 74 imply Phred + 64) [8].

2.2.2 Alignment Formats: SAM, BAM, and CRAM

Once reads are mapped to a reference genome, the data is stored in the sequence alignment/Map (SAM) format or its binary derivatives. This format is the bedrock of variant calling and downstream analysis [6]. SAM is a human-readable, tab-delimited text format consisting of a header section (metadata) and an alignment section.

- **The FLAG Field**: The second column is a bitwise integer that encodes the alignment status of the read. It is constructed by summing binary flags.
  - $1(2^0)$: Read is paired.
  - $2(2^1)$: Read is mapped in a proper pair.
  - $4(2^2)$: Read is unmapped.
  - $16(2^4)$: Read is on the reverse strand.
  - $1024(2^{10})$: Read is a PCR or optical duplicate.
  - For example, a flag of 99 (binary 01100011) indicates: paired + proper pair + mate reverse strand + first in pair. Understanding these flags is essential for filtering; removing duplicates (− F 1024) is a standard cleaning step [9, 10].

- **The CIGAR String**: The "compact idiosyncratic gapped alignment report" describes the detailed alignment structure. 100M denotes 100 bases matching or mismatching the reference. 50M2I48M denotes 50 matches, a 2-based insertion in the read, and 48 matches. This string allows the reconstruction of the read's placement relative to the reference without storing the full sequence redundancy [11].
- **Binary Alignment/Map (BAM)**: BAM is the binary equivalent of SAM. It utilizes Blocked GNU Zip Format (BGZF) compression. Unlike standard gzip, BGZF compresses data in independent blocks, allowing for random access. An accompanying index file (.bai) maps genomic coordinates to file offsets, enabling tools to jump instantly to "Chr1:1,000,000" without reading the entire file. This random-access capability is crucial for genome browsers and parallel processing [12].
- **Compressed Reference-Oriented Alignment Map (CRAM)**: As datasets grew to petabyte scales, the redundancy of storing millions of reads identical to the reference genome became untenable. CRAM solves this by reference-based compression.
  - **Delta Encoding**: CRAM stores only the differences between a read and the reference. If a read matches the reference perfectly, the sequence is not stored at all.
  - **Lossy Compression Options**: CRAM supports "lossy" modes, primarily for quality scores. Since many analyses do not require the distinction between ***Q38* and *Q39***, CRAM can "bin" these scores (e.g., all scores 35–39 become 37), drastically reducing file size with negligible impact on variant calling accuracy.
  - **Independence**: While originally dependent on the reference file for reading, modern CRAM specifications support embedding the reference checksums or sequences to prevent "broken" files if the reference is lost [13].

2.2.3 Variant Data: The VCF

The variant call format (VCF) stores the output of variant callers, the sites where the sample differs from the reference [6].

- **Structure and Fields**: A VCF file contains meta-information lines (starting with ##), a header line, and data lines.
  - **INFO Field**: A semicolon-separated list of key-value pairs describing the variant site itself. Common tags include DP (total depth), AF (allele frequency), and MQ (RMS mapping quality).
  - **FORMAR and Sample Fields**: These columns describe the genotype for each sample.

    Genotype (GT): 0/0 (homozygous reference), 0/1 (heterozygous), 1/1 (homozygous alternate).

Table 2.1 Features of interval formats

Feature	Browser extensible data (BED)	GFF/GTF (general feature format)
Origin	UCSC Genome Browser	Sanger/Ensembl
Start coordinate	**0-based** (first base is 0)	**1-based** (first base is 1)
End coordinate	**Exclusive** (half-open)	**Inclusive** (closed)
Interval logic	"	
Example (first 100 bases)	Chrom 0 100	Chrom source type 1 100
Length calculation	End–Start	End–Start + 1

Phred-Scaled Likelihoods (PL): The probabilistic heart of the VCF. PL is a list of three numbers (for diploid) representing the probability that the true genotype is 0/0, 0/1, or 1/1, scaled by $-\mathbf{10\log_{10}}$. A PL of 0, 30, 50 means the 0/0 genotype is most likely ($10^{-0} = 1$), while 0/1 is 10^{-3} less likely. This probabilistic retention is vital for joint genotyping cohorts [14].

2.2.4 Interval Formats: BED Versus GFF/GTF

Genomic features (genes, peaks, regulatory regions) are defined by intervals. A frequent source of "off-by-one" errors in bioinformatics stems from the differing coordinate systems of these formats. The information of various features of interval formats is in Table 2.1.

- **Implications**: When converting between formats or querying databases, failure to adjust coordinates (adding or subtracting 1) can lead to frameshifts in gene models or extracting the wrong sequence for motif analysis [15].

2.3 Data Cleaning, Normalization and Quality Control

The transition from raw signal to interpretable data involves a rigorous purification process. Raw sequencing data is inherently noisy, laden with technical artifacts such as adapter contamination, PCR duplicates, and batch effects. Without stringent quality control (QC) and normalization, these artifacts can easily mimic or mask true biological signals.

2.3.1 Quality Control and Trimming

The first line of defense in any pipeline is the assessment of raw read quality. The quality metrics and tools are as follows:

- **FastQC**: This tool is the industry standard for visualizing quality metrics. It performs a series of modular analyses [16]:
  - **Per Base Sequence Quality**: Visualizes the distribution of Phred scores across the read length. A characteristic drop in quality at the 3' end of sequencing reads is common due to enzyme fatigue and dephasing.
  - **Per Base Sequence Content**: Checks for biases in A/C/G/T proportions. In a random library, these should be parallel; divergences at the start of reading often indicate non-random priming biases (e.g., in RNA-seq).
  - **Overrepresented Sequences**: Identifies sequences that appear more frequently than expected by chance, which are usually flag adapter contamination or PCR primers.
- **MultiQC**: In large-scale projects involving hundreds of samples, inspecting individual FastQC reports is unfeasible. MultiQC parses the log files from FastQC, aligners (STAR, BWA), and quantifiers (Salmon) to generate a single, interactive HTML report. This aggregate view is critical for identifying outlier samples and systematic batch effects across a cohort [17].
- **Trimming Algorithms**: Preprocessing involves the algorithmic removal of low-quality bases and technical sequences.
  - **Trimmomatic**: A robust, Java-based tool known for its "sliding window" approach. It scans the read from the 5' to 3' end, calculating the average quality in a small window (e.g., 4 bases). If the average drops below a threshold (e.g., Q20), the reader is clipped at that point. Trimmomatic is particularly adept at handling paired-end data, ensuring that if one read is discarded, its mate is also removed or flagged as "unpaired" to maintain file synchronization [18].
  - **Cutadapt**: While capable of quality trimming, cutadapt is primarily specialized for adapter removal. It uses semi-global alignment to find adapter sequences even in the presence of sequencing errors, making it highly flexible for custom protocols [19].
  - **fastp**: A modern, all-in-one C++preprocessor that has gained popularity for its speed and functionality. Unlike the two-step process of FastQC followed by Trimmomatic, fastp performs quality profiling, adapter trimming, poly-G tail trimming (essential for two-color chemistry sequencers like NextSeq/NovaSeq), and filtering in a single pass. Benchmarks demonstrate it is 2–5 times faster than traditional tools while providing equivalent or superior accuracy [18].

2.3.2 RNA-Seq Normalization

In transcriptomics, the raw number of reads mapping to a gene is not a direct measure of expression. It is confounded by two primary technical factors: Sequencing depth (library size) and gene length.

- **Depth Bias**: A sample sequenced to 50 million reads will have roughly double the counts of a sample sequenced to 25 million reads, assuming equal expression.
- **Length Bias**: In standard fragmentation-based protocols, a longer transcript produces more fragments (and thus more reads) than a shorter transcript at the same molar concentration.

To compare expression levels, these biases must be mathematically corrected.

- **Within-Sample Normalization**: When the goal is to compare Gene A to Gene B within the same sample (e.g., "In actin expressed higher than tubulin?"), we must correct for gene length.
  - **Transcripts Per Million (TPM)**: The calculation proceeds in two steps:
    Normalize for Length: Divide the raw count (C_i) by the gene length (L_i) in kilobases to get Reads Per Kilobase (RPK_{i}).

$$RPK_i = \frac{C_i}{L_i}$$

    Normalize for Depth: Sum all RPK values in the sample to get a scaling factor (S). Divide each RPK_i by S and multiply by 10^6.

$$TPM_i = \frac{RR_i}{\sum PRK_j} \times 10^6$$

    Key Property: The sum of TPM values in any sample is always 1,000,000. This invariant property makes TPMs more intuitive and comparable than the older RPKM/FPKM metrics, which do not sum to a constant and can be skewed by mean library size changes [20].
- **Between Sample Normalization: TMM and Size Factors**
  When the goal is differential expression (DE) analysis (comparing Gene A in Sample 1 vs. Sample 2), TPM is not ideal. DE tools like DESeq2 and edgeR expect raw counts to model the mean-variance relationship (noise) using the negative binomial distribution. Furthermore, TPM is sensitive to compositional bias: if a few highly expressed genes (e.g., hemoglobin in blood) take up 50% of the reads in Sample 1 but only 10% in Sample 2, the TPMs of all other genes in Sample 1 will be artificially compressed, appearing downregulated.

- **Trimmed Mean of M-values (TMM)**: Implemented in the edgeR package. TMM assumes that most genes are not differentially expressed. It selects a reference sample and calculates the log-ratios (M-values) of expression for every gene relative to the reference. It then trims the genes with the highest and lowest expression (A-values) and the highest and lowest log-rations (M-values). The weighted mean of the remaining M-values is used as the normalization factor. This makes TMM extremely robust to outliers and compositional differences [20].
- **DESeq2 (Median of Ratios)**: Similarly, DESeq2 calculates a pseudo-reference sample (geometric meaning of each gene across all samples). For each sample, it computes the ratio of a gene's count to the pseudo-reference. The median of these ratios is the size factor for that sample. The median is robust to extreme outliers, ensuring that a few highly differentially expressed genes do not skew the normalization of the entire library [21].
- **Summary Recommendation**: Use TPM for visualization (heatmaps), clustering, and within-sample comparisons. Use TMM/size factors (via edgeR/DESeq2) on raw counts for statistical differential expression analysis.

2.4 Feature Representation

The raw strings of A, C, G, and T that comprise genomic data are not inherently suitable for modern machine learning algorithms, which require numerical vectors as input. The transformation of sequence data into numerical features—feature representation—is a critical step that determines the ability of a model to learn biological patterns.

2.4.1 K-mer Representations

The simplest and most pervasive method of representation is the decomposition of a sequence into substrings of length k, known as **K-mers.**

- **Bag of K-mers**: Analogous to the "Bag of Words" model in Nature Language Processing (NLP), a sequence is represented as a frequency vector of its constituent k-mers. For a standard alphabet size of 4, a k-mer size of 3 yields a vector of dimension $4^3 = 64$.
- **Applications**: This representation is foundational for alignment-free comparison methods. In metagenomics, species identification tools use k-mer profiles to bin contigs, relying on the fact that different genomes possess distinct k-mer signatures (e.g., GC-rich organisms have different profiles than AT-rich ones) [2].
- **One-Hot Encoding**: To preserve positional information (which "Bag of k-mer" discards), sequences are often One-Hot encoded. A sequence of length L becomes a matrix of size $4 \times L$, where each column has a single "1" corresponding to the

nucleotide at that position. While simple, this representation is sparse and high dimensional, often leading to the "curse of dimensionality" in deep learning models [22].

2.4.2 Genomic Embeddings: DNA2Vec

To overcome the limitations of sparse representations, bioinformatics has adapted the concept of Word Embeddings from NLP. The hypothesis is that k-mers, like words, derive meaning from their context.

- **DNA2Vec**: This algorithm applies the Word2Vec (continuous bag-of-words or skip-gram) architecture to genomic data. It trains a shallow neutral network on massive corpora of DNA sequences to predict a k-mer given to its neighbors.
  - **Result**: The network learns a dense, continuous vector (embedding) for each k-mer. Crucially, these vectors capture biological similarity. The cosine similarity between the vectors of two k-mers correlates with their sequence similarity (Needleman–Wunsch score).
  - **Vector Arithmetic**: DNA2Vec vectors exhibit additive properties were summing the vectors of k-mers approximates the vector of their concatenation. This allows variable-length sequences to be represented by summing their constituent k-mer embeddings [23].
  - **Graph-Based Embeddings (K-mer-Node2Vec)**: Standard DNA2Vec can be slow to train on large genomes. K-mer-Node2Vec constructs a k-mer co-occurrence graph and uses random walks to sample contexts, significantly accelerating the training process while maintaining embedding quality [23].

2.4.3 Genomic Foundation Models

The success of transformer models like GPT and BERT in language has spurred the development of genomic foundation models. These are large deep learning models pre-trained on vast amounts of genomic data to learn the syntax and grammar of DNA.

- **DNABERT**: Based on the BERT architecture, this model is trained using Masked Language Modeling (MLM). Random k-mers in a sequence are masked, and the model must predict them based on the surrounding context. Through this process, DNABERT learns to recognize complex regulatory motifs, such as promoter elements and splice sites, without explicit supervision. DNABERT-2 has optimized this approach, achieving state-of-the-art performance on tasks like transcription factor binding prediction [24].
- **Nucleotide Transformer**: This model scales up the concept, training on 25,000 diverse genomes. By learning from multi-species data, it captures universal biological constraints. A key application is Zero-Shot Scoring: the model can predict

the pathogenicity of a variant (e.g., a SNP) by calculating how much the mutation decreases the probability (likelihood) of the sequence compared to the reference, effectively quantifying "unnaturalness" without needing a labeled disease dataset [24].
- **HyenaDNA**: Standard transformers scale quadratically with sequence length $O(N^2)$, limiting their context window. HyenaDNA replaces the attention mechanism with implicit convolutions, scaling sub-quadratically $O(N \log N)$. This allows it to process context windows of up to 1 million base pairs, enabling the model to learn extremely long-rand interactions, such as those between distal enhancers and promoters [25].

2.4.4 Graph Genomes

The traditional representation of a genome is linear (a string of bases). This fails to capture population diversity, leading to reference bias: reading from a sample that differs significantly from the reference (e.g., containing a large insertion) will fail to align.

- **Variation Graphs (Pangenomes)**: These data structures represent the genome as a graph.
  - **Nodes**: Segments of DNA sequence.
  - **Edges**: Connections between segments.
  - **Paths**: An individual genome is a specific path walking through the graph.
- **Structure**: Variants appear as "bubbles" (alternative paths) in the graph. By mapping reads to a variation graph (using tools like vg), aligners can traverse these bubbles, correctly placing reads that would otherwise be unmapped or misaligned. This is critical for hypervariable regions like the Major Histocompatibility Complex (MHC) [26].

2.5 High-Performance and Cloud Computing

The scale of genomic data, where a single whole genome sequencing (WGS) sample can be 100 GB and population cohorts reach petabytes, renders desktop computing obsolete. Modern genomics relies on high-performance computing (HPC) and increasingly, cloud infrastructure.

2.5.1 Containerization: The Unit of Reproducibility

A major crisis in bioinformatics is the fragility of software environments. A pipeline that runs on one cluster often fails on another due to differences in system libraries (e.g., glibc versions) or tool dependencies.

- **Docker**: The industry standard for containerization. It packages code, libraries, and the operating system into an immutable image. However, Docker requires a root-privileged daemon to run, which poses a severe security risk in multi-user HPC environments (a user could mount the host filesystem and delete root files) [27].
- **Singularity (Apptainer)**: Designed specifically for scientific HPC. It runs containers in user-space without root privileges. Singularity can import Docker images and convert them into a secure file format (.sif). It also integrates natively with message passing interface (MPI) for parallel jobs, making it the preferred container runtime for supercomputers [27].

2.5.2 Cloud Architectures (AWS, GCP)

Cloud computing offers "elastic" scalability, the ability to provision 10,000 CPUs for a few hours and then release them. This fits the "bursty" nature of genomic analysis perfectly.

- **Data Lakes and Storage Tiering**: Genomic data has a distinct lifecycle, high access during analysis, followed by long periods of dormancy. Cloud object storage (like AWS S3) supports tiering to optimize costs.
  - **Hot Storage (S3 Standard)**: For active processing, high cost, low latency.
  - **Warm Storage (S3 Infrequent Access)**: For BAM/CRAM files that might be accessed once a month. Lower storage cost, retrieval fees apply.
  - **Cold Storage (Glacier Deep Archive)**: For raw FASTQ files post-analysis. Extremely low cost (fractions of a cent per GB), but retrieval takes 12–48 h. Designing lifecycle policies to automatically move data to colder tiers is essential for budget management [28].
- **Spot Instances and Cost Optimization**: A major cost driver in the cloud is computing.
  - **Spot/Preemptible Instances**: Cloud providers (AWS, Google) sell excess capacity at steep discounts (60–90% off on-demand prices). The caveat is Preemption, the provider can reclaim the instance with very short warning (e.g., 2 min on AWS, 30 s on GCP) if they need the capacity back.
  - **Strategies for Genomics**:

    Fault Tolerance: Workflow managers like Nextflow can automatically detect a Spot interruption and resubmit the job (potentially to a stable on-demand instance).

    Checkpointing: Advanced tools like MemVerge Memory Machine use technologies like CRIU (Checkpoint/Restore in Userspace) to "freeze" the RAM state of a running job and save it to disk. If a spot instance is preempted, the job can be restored on a new instance from the checkpoint, rather than restarting from the beginning. This makes it feasible to run long-running jobs (e.g., assembly) on volatile Spot instances [29].

Real-World Impact: The broad institute optimized their GATK pipeline to run on GCP preemptible VMs, reducing the computer cost of a whole genome to roughly $5 [30].

2.6 Data Integration

Biological insight increasingly comes not from a single assay, but from the synthesis of multiple data types. Data integration techniques aim to merge these disparate signals while removing technical noise.

2.6.1 Horizontal Integration: Batch Correction

"Horizontal" integration involves combining datasets of the same type (e.g., scRNA-seq) from different sources (batches). The major challenge is batch effect: technical differences (e.g., different sequencing lanes, reagents, or labs) often dominate biological variation.

- **Harmony**: A leading algorithm for single-cell integration. It projects cells into a shared PCA space and then calculates an iterative correction. It pulls cells with similar biological profiles (clusters) from different batches together, while pushing dissimilar cells apart. It is favored for its speed and ability to scale to millions of cells [31].
- **Seurat v5 Integration**: Introduces a flexible framework allowing users to choose integration anchors (using CCA—Canonical Correlation Analysis, or RPCA). It introduces bridge integration, which allows the mapping of a multi-omic dataset (e.g., ATAC + RNA) to a unimodal reference (e.g., a massive RNA-only atlas) by using a "dictionary" of cells to translate between modalities [32].

2.6.2 Vertical Integration: Multi-Omics Factor Analysis

"Vertical" integration involves combining different data types (views) from the same samples.

- **MOFA + (Multi-omics Factor Analysis)**: This unsupervised method generalizes principal component analysis (PCA) to multiple matrices. It decomposes the data into a set of Laten Factors (Z) and Weight (W).
  - Decomposition: $Data \simeq Factors \times Weight$
  - Insight: MOFA + determines how variance each factor explains in each view.

Shared Factors: A factor explaining variance in both Methylation and Gene Expression might represent a biological driver (e.g., a transcription factor).
Private Factors: A factor explaining variance only in RNA-seq might represent technical noise specific to that assay (e.g., RNA degradation).
  - Robustness: MOFA + handles missing data gracefully; if a sample is missing proteomics data, it can still contribute to the model based on its genomics/transcriptomics data [33].

2.6.3 Pangenomics Construction

Constructing a pangenome is the ultimate data integration challenge merging complete genomes from diverse populations.

- **Minigraph-Cactus**: This pipeline represents state-of-the-art.
  - **Minigraph**: constructs the initial graph structure using structural variants (SVs). It focuses on large-scale topology.
  - **Cactus**: A base-level aligner that fills in the details, mapping small variants (SNPs/Indels) onto the graph structure.
  - **Result**: A graph that captures both large structural diversity and fine-grained sequence variation. This graph serves as a superior reference for mapping, significantly reducing errors in regions that are highly polymorphic in the human population [34].

2.7 Scalable Genomic Data Pipelines

The complexity of modern genomic analysis, often involving pipelines with 50 + distinct steps and terabytes of intermediate data, has made manual execution via shell scripts untenable. Scalability, reproducibility, and portability are now managed by specialized workflow management systems (WfMS).

2.7.1 Workflow Management Systems (WfMS)

Table 2.2 gives the comparison of major genomic workflow managers.

- **Nextflow**: Built on the dataflow programming model. Processes are connected via "channels." As soon as a piece of data is emitted into a channel, the downstream process triggers automatically. This abstraction allows Nextflow to handle massive parallelism implicitly. Its strict separation of configuration (infrastructure) from logic

Table 2.2 Comparison of major genomic workflow managers

Feature	Nextflow	Snakemake	WDL (Cromwell)
Language base	Groovy (Java)	Python	Domain Specific (Declarative)
Execution model	Dataflow (push-based)	Dependency graph (pull-based)	Scatter–Gather
Containerization	Native (Docker/ Singularity/Podman)	Native (Conda/ Singularity)	Native (Docker)
Cloud support	AWS Batch, Google Batch, K8s, Azure	AWS (via plugins), Google, K8s	Google Life Sciences (Terra), AWS
Primary advantage	Portability and Implicit Parallelism	Python Integration and Ease of Learning	Industry Standard (Broad/GATK)

(pipeline code) means the exact same script can run on a laptop or a Kubernetes cluster by simply switching a config file [35].

- **Snakemake**: Inspired by GNU make, it builds a dependency graph. The user requests a final output file, and Snakemake works backwards to determine which rules must be executed. Its deep integration with Python allows users to write arbitrary Python code (e.g., pandas for sample sheet parsing) directly within the workflow file, making it highly flexible for developers comfortable with Python [35].
- **Workflow Description Language (WDL)**: Developed by the Broad Institute, WDL is a human-readable, declarative language designed to describe tasks and workflows. It is the language of the GATK best practices and is the native format for the Terra cloud platform. WDL separates the description of the workflow from the execution engine (Cromwell), facilitating sharing across different infrastructures [36].

2.7.2 Case Study: The GATK Best Practices Pipeline

To illustrate the orchestration required, consider the GATK Germline Variant Discovery pipeline, which is the industry standard for identifying SNPs and Indels in WGS data [37].

- **Data Preprocessing**:
  - **Mapping**: Raw FASTQ reads are aligned to the reference (GRCh38) using BWA-MEM.
  - **Mark Duplicates**: (Picard) Identifies and tags reads originating from the same DNA fragment (PCR artifacts).
  - **Base Quality Score Recalibration (BQSR)**: A machine learning step that builds an empirical error model. It detects systematic errors (e.g., a specific sequencer cycle

consistently overcalling "A") and adjusts the Phred scores in the BAM file to match the empirical error rate.

- **Variant Calling (HaplotypeCaller)**:
  - Unlike simple pileup callers, HaplotypeCaller performs local de novo assembly of reads in active regions. It builds a graph to reconstruct the likely haplotypes and calculates the likelihood of each genotype.
  - **Output**: A gVCF (Genomic VCF). Unlike a standard VCF, a gVCF contains blocks of records for non-variant sites, providing confidence that a site is homozygous reference.
- **Joint Genotyping**:
  - **GenomicsDBImport**: Consolidates gVCFs from hundreds of samples into a highly optimized database (TileDB).
  - **GenotypeGVCFs**: Uses the population wide data to resolve ambiguous calls in individuals (e.g., a low-coverage site in one person might be rescued if the variant is seen clearly in others).
- **Variant Filtering (VQSR/CNN)**:
  - Hard filters (e.g., "Quality < 30") are brittle. Variant Quality Score Recalibration (VQSR) uses a Gaussian Mixture Model trained on "truth sets" (e.g., HapMap, Omni, 1000G) to learn the multidimensional profile of real variants versus artifacts (strand bias, mapping quality, etc.) and assigns a probability score (VQSLOD) to every call [37].

This pipeline exemplifies the themes of this chapter: the need for precise data formatting (BAM/gVCF), rigorous statistical modeling (BQSR/VQSR), and scalable execution (WDL/Nextflow) to transform raw nucleotide sequences into reliable biological knowledge.

2.8 Summary

This chapter provides foundational aspects of genomic data and computational methods essential for modern genomic analysis. It highlights the shift in biology toward a data-driven discipline, driven by advancements in nucleotide sequencing technologies. The chapter categorizes genomic data into DNA sequencing, transcriptomics, epigenomics, metagenomics, and multi-omics, explaining their unique properties, applications, and challenges. It discusses the technical specifications of key data formats like FASTQ, SAM/BAM/CRAM, VCF, and interval formats, emphasizing their structure, encoding, and importance for data interoperability. The chapter also covers data cleaning, normalization, and quality control processes, including tools like FastQC, MultiQC, and trimming algorithms. It explores feature representation methods such as k-mer representations, DNA2Vec embeddings, and genomic foundation models like DNABERT and HyenaDNA. Additionally, it introduces advanced concepts like pangenomics and graph genomes to address population diversity and reference bias. The chapter concludes with insights into scalable genomic data pipelines, workflow management systems like Nextflow, Snakemake, and WDL, and the role of high-performance and cloud computing in handling large-scale genomic datasets.

References

1. Williams, G. A., Liede, S., Fahy, N., Aittomaki, K., Perola, M., Helander, T., ... & Sagan, A. (2021). *Regulating the unknown: A guide to regulating genomics for health policy-makers.* https://www.ncbi.nlm.nih.gov/books/NBK569502/
2. Graw, S., Chappell, K., Washam, C. L., Gies, A., Bird, J., Robeson, M. S., & Byrum, S. D. (2021). Multi-omics data integration considerations and study design for biological systems and disease. *Molecular Omics, 17*(2), 170–185.
3. Faux, T., Rytkönen, K. T., Mahmoudian, M., Paulin, N., Junttila, S., Laiho, A., & Elo, L. L. (2021). Differential ATAC-seq and ChIP-seq peak detection using ROTS. *NAR Genomics and Bioinformatics, 3*(3), lqab059.
4. ChIP-seq vs. ATAC-seq. https://www.cd-genomics.com/resource-chip-seq-vs-atac-seq.html. Last Access: January 2026.
5. Garcia, K. (2026). *Integrated ATAC, ChIP, HiC and RNA sequencing shed light on the chromatin interactions puzzle.* https://medium.com/@cd-genomics/integrated-atac-chip-hic-and-rna-sequencing-shed-light-on-the-chromatin-interactions-puzzle-20b8acbd1a7c. Last Access: January 2026.
6. Overview of Different Data Formats. https://www.futurelearn.com/info/courses/making-sense-of-genomic-data-covid-19-web-based-bioinformatics/0/steps/319350. Last Access: January 2026.
7. Common File Formats in Bioinformatics. https://docs.genebe.net/docs/handbook/file-formats/. Last Access: January 2026.
8. Quality Score Encoding—Illumina Support. https://support.illumina.com/help/BaseSpace_Sequence_Hub_OLH_009008_2/Source/Informatics/BS/QualityScoreEncoding_swBS.htm. Last Access: January 2026.
9. NGS File Formats—Bioinformatics. https://bioinformatics.ccr.cancer.gov/class_manager/file/class_file/view/5fb5cc2cc576e6cf8071844c982b4d84. Last Access: January 2026.
10. Decoding SAM Flags. https://broadinstitute.github.io/picard/explain-flags.html. Last Access: January 2026.
11. Sequence Alignment/Map Format Specification—Samtools. https://samtools.github.io/hts-specs/SAMv1.pdf. Last Access: January 2026.
12. Li, H., Handsaker, B., Wysoker, A., Fennell, T., Ruan, J., Homer, N., ... & 1000 Genome Project Data Processing Subgroup. (2009). The sequence alignment/map format and SAMtools. *Bioinformatics, 25*(16), 2078–2079.
13. About Alignment Files (BAM and CRAM). https://www.internationalgenome.org/faq/about-alignment-files-bam-and-cram/. Last Access: January 2026.
14. What Is a Variant Call Format (VCF) File? https://genomoncology.com/blog/what-is-a-variant-call-format-vcf-file/. Last Access: January 2026.
15. Slowikowski, K. (2013). *0-based and 1-based genomic intervals, overlap, and distance.* https://slowkow.com/notes/genomic-intervals/. Last Access: January 2026.

16. Andersson, R., Gebhard, C., Miguel-Escalada, I., Hoof, I., Bornholdt, J., Boyd, M., ... & Andrews, S. (2010). *FastQC: A quality control tool for high throughput sequence data.* Version 0.11, 2.
17. Using MultiQC Reports. https://docs.seqera.io/multiqc/reports. Last Access: January 2026.
18. Chen, S. (2025). fastp 1.0: An ultra-fast all-round tool for FASTQ data quality control and preprocessing. *Imeta, 4*(5), e70078.
19. Chen, S., Zhou, Y., Chen, Y., & Gu, J. (2018). Fastp: An ultra-fast all-in-one FASTQ preprocessor. *Bioinformatics, 34*(17), i884–i890.
20. Lüleci, H. B., Uzuner, D., Cesur, M. F., İlgün, A., Düz, E., Abdik, E., ... & Çakır, T. (2024). A benchmark of RNA-seq data normalization methods for transcriptome mapping on human genome-scale metabolic networks. *npj Systems Biology and Applications, 10*(1), 124.
21. Pluto. (2023). *Navigating RNA-Seq data: A comprehensive guide to normalization methods.* https://pluto.bio/resources/Learning%20Series/navigating-rna-seq-data-a-guide-to-normalization-methods. Last Access: January 2026.
22. Ng, Patrick. *dna2vec: Consistent vector representations of variable-length k-mers.* arXiv preprint arXiv:1701.06279 (2017).
23. Yu, Z., Yang, Z., Lan, Q., Wang, Y., Huang, F., & Cai, Y. (2023, July). Kmer-node2vec: A fast and efficient method for kmer embedding from the kmer co-occurrence graph, with applications to DNA sequences. In *2023 45th Annual International Conference of the IEEE Engineering in Medicine & Biology Society (EMBC)* (pp. 1–4). IEEE.
24. Guo, F., Guan, R., Li, Y., Liu, Q., Wang, X., Yang, C., & Wang, J. (2025). Foundation models in bioinformatics. *National science review, 12*(4), nwaf028.
25. Feng, H., Wu, L., Zhao, B., Huff, C., Zhang, J., Wu, J., ... & Wu, C. (2024). Benchmarking DNA foundation models for genomic sequence classification. *bioRxiv.*
26. Veteam/vg: tools for working with genome variation graphs. https://github.com/vgteam/vg. Last Access: January 2026.
27. Hughes, A. (2023). *An introduction to singularity containers.* https://sylabs.io/2023/03/an-introduction-to-singularity-containers/. Last Access: January 2026.
28. Transferring Genomics Data to the Cloud and Establishing Data Access Patterns Using AWS DataSync and AWS Storage Gateway for Files. https://docs.aws.amazon.com/whitepapers/latest/genomics-data-transfer-analytics-and-machine-learning/transferring-genomics-data-to-the-cloud-and-establishing-data-access-patterns-using-aws-datasync-and-aws-storage-gateway-for-files.html. Last Access: January 2026.
29. Paulau, G., & Kniep, C. (2025). *Maximizing EC2 spot instance reliability for Nextflow on AWS batch with Memorh machine batch.* https://aws.amazon.com/blogs/publicsector/maximizing-ec2-spot-instance-reliability-for-nextflow-on-aws-batch-with-memory-machine-batch/. Last Access: January 2026.
30. Hajian, A. (2025). *Understanding google cloud costs—Sample use cases.* https://support.terra.bio/hc/en-us/articles/360029772212-Understanding-Google-Cloud-costs-sample-use-cases. Last Access: January 2026.
31. Tran, H. T. N., Ang, K. S., Chevrier, M., Zhang, X., Lee, N. Y. S., Goh, M., & Chen, J. (2020). A benchmark of batch-effect correction methods for single-cell RNA sequencing data. *Genome Biology, 21*(1), 12.
32. Hastings, J., Lee, D., & O'Connell, M. J. (2024). Batch-effect correction in single-cell RNA sequencing data using JIVE. *Bioinformatics Advances, 4*(1), vbae134.
33. Argelaguet, R., Velten, B., Arnol, D., Dietrich, S., Zenz, T., Marioni, J. C., ... & Stegle, O. (2018). Multi-Omics Factor Analysis—A framework for unsupervised integration of multi-omics data sets. *Molecular Systems Biology, 14*(6), MSB178124.

34. Hickey, G., Monlong, J., Ebler, J., Novak, A. M., Eizenga, J. M., Gao, Y., … & Paten, B. (2024). Pangenome graph construction from genome alignments with Minigraph-Cactus. *Nature Biotechnology, 42*(4), 663–673.
35. Langer, B. E., Amaral, A., Baudement, M. O., Bonath, F., Charles, M., Chitneedi, P. K., … & Community, T. N. C. (2025). Empowering bioinformatics communities with Nextflow and nf-core. *Genome Biology, 26*(1), 228.
36. Bioinformatics Pipeline Framework. (2025). *Nextflow vs Flyte vs Airflow vs Snakemake.* https://www.tracer.cloud/resources/bioinformatics-pipeline-frameworks-2025. Last Access: January 2026.
37. GATK Team. (2024). *Germline short variant discovery (SNPs + Indels).* https://gatk.broadinstitute.org/hc/en-us/articles/360035535932-Germline-short-variant-discovery-SNPs-Indels. Last Access: January 2026.

AI Methods in Genomic Analysis 3

Advances in high-throughput sequencing (HTS) have transformed genomics into one of the most data-intensive domains in life sciences. Modern sequencing platforms now generate massive datasets spanning whole-genome, transcriptome, epigenome, and metagenome profiling, alongside population-scale genomic surveys involving thousands to millions of individuals. These datasets are characterized by extreme dimensionality, high sparsity, complex correlation structures, and subtle signal-to-noise ratios, posing significant challenges to conventional statistical and rule-based bioinformatics methods [1, 2]. In parallel, the emergence of planetary-scale k-mer indexing and pangenomic resources has further increased both computational demands and analytical opportunities, enabling large-scale comparisons across diverse genomes and facilitating population genetics, evolutionary inference, and ancestry-aware analyses [3, 4].

Traditional bioinformatics pipelines remain indispensable for foundational tasks such as sequence alignment, genome assembly, variant calling, and functional annotation. These pipelines are typically optimized for deterministic processing and well-defined biological assumptions, which ensure robustness and reproducibility in routine analyses. However, they often struggle to capture nonlinear interactions, epistatic relationships, and long-range regulatory dependencies that are increasingly recognized as central to genome function, particularly when integrating multiple omics layers or exploring combinatorial effects of rare variants [1, 2]. For example, k-mer-based approaches are highly efficient for sequence matching and genome indexing, yet their application to large, heterogeneous datasets requires careful parameterization and substantial computational infrastructure, limiting their flexibility in integrative or exploratory analyses at scale [3]. Similarly, combining genomic, transcriptomic, epigenomic, and clinical data in contexts such as precision oncology or large-scale screening requires computational frameworks capable

K. Shaikh et al., *Artificial Intelligence in Genomics*, Synthesis Lectures on Engineering, Science, and Technology, https://doi.org/10.1007/978-3-032-25486-3_3

Table 3.1 Key 2025 trends linking AI and genomics

Dimension	Recent emphasis in 2025 literature	Citations
Data scale and complexity	HTS, multi-omics, planetary-scale k-mers, pangenomes	[1–4]
Core AI roles	Sequence classification, functional annotation, multi-omics integration, drug discovery	[1, 2, 5]
Emerging paradigms	Foundation models, generative AI, hybrid quantum–classical AI, data governance frameworks	[1, 2, 4, 5]

of modeling subtle cross-modal dependencies while adhering to strict data governance and privacy constraints [1, 4].

Artificial intelligence (AI) has emerged as a pivotal paradigm for addressing these methodological limitations. Classical machine learning techniques, including random forests and support vector machines, have been widely applied to tasks such as sequence classification, functional annotation, and variant prioritization, offering improved performance over linear models through their ability to capture nonlinear decision boundaries. More recently, deep learning architectures, particularly convolutional and transformer-based models, have demonstrated the capacity to process long genomic sequences and heterogeneous inputs, learning hierarchical representations that reflect biologically meaningful patterns such as motifs, regulatory syntax, and chromatin-associated signals [1, 2, 5]. Transformer-based foundation models enable pretraining on large-scale genomic corpora and transfer of learned representations across tasks, reducing data requirements and improving generalization in downstream analyses [5]. Generative AI approaches are also gaining traction, providing computational frameworks for simulating realistic genomic sequences, probing sequence–function relationships, and guiding the design of proteins or regulatory elements under biologically constrained settings [1, 5]. When coupled with biologically informed feature representations—such as k-mer embeddings, sequence motifs, or structural annotations—these models can offer both predictive power and mechanistic insight.

Despite these advances, the responsible and effective application of AI in genomics requires sustained attention to data stewardship, bias mitigation, and interpretability. Genomic datasets often reflect population, technical, or sampling biases that can be amplified by data-driven models if not carefully addressed. Ensuring that AI systems are robust, reproducible, and biologically meaningful is therefore essential for translational impact, whether in guiding experimental design, prioritizing candidate therapeutic targets, or informing personalized medicine strategies [1, 4]. Transparent model evaluation, curated training data, and interpretable outputs remain central requirements for the integration of AI into genomic research and practice.

Table 3.1 summarizes key 2025 trends linking AI and genomics, highlighting areas of emphasis in data scale, AI roles, and emerging methodological paradigms.

In this chapter, we provide a structured overview of AI methodologies for genomic analysis, progressing from classical machine learning approaches to deep learning, transformer-based foundation models, and generative AI. Emphasis is placed on methodological principles, feature representations, biological relevance, and translational potential, while critically addressing challenges related to interpretability, data governance, and scalable computation in modern genomic research.

3.1 Classical Machine Learning Methods in Genomics

Classical machine learning (ML) methods constitute a foundational category of data-driven tools in genomic analysis, offering structured and mathematically well-defined approaches for inferring associations between high-dimensional genomic features and biologically relevant outcomes. Unlike contemporary deep learning models that autonomously learn hierarchical representations from raw data, classical ML algorithms typically operate on engineered features and rely on explicit optimization objectives. This design confers advantages in computational efficiency, transparency, and interpretability—properties that are particularly valuable in genomic contexts where data dimensionality is high, sample sizes are limited, and model outputs must often be scrutinized for biological plausibility or downstream experimental validation [6, 7]. Genomic datasets commonly exhibit a "large p, small n" structure, in which the number of measured features—such as single-nucleotide polymorphisms (SNPs), gene expression levels, or epigenetic marks—far exceeds the number of available samples. Classical ML methods are well suited to this regime when combined with appropriate preprocessing steps, including normalization, dimensionality reduction, feature selection, and regularization. As a result, these models remain integral to genomic research not only as interpretable predictors but also as baseline benchmarks and complementary components within larger analytic pipelines that may incorporate more complex AI methodologies. This section reviews the principal classes of classical machine learning methods used in genomics, emphasizing their conceptual mechanics, typical applications, and biological relevance. Particular attention is given to how their modeling assumptions shape interpretability and performance, thereby motivating the transition to more expressive learning paradigms discussed in subsequent sections.

3.1.1 Linear and Regularized Regression Techniques for Genomic Prediction

Logistic regression is a probabilistic classification model that estimates the log-odds of discrete outcomes as a linear function of input predictors. In genomic analysis, its coefficients provide directly interpretable estimates of the contribution of individual loci, genes, or molecular features to a phenotype. Such transparency allows researchers to link statistical associations to known biological pathways, regulatory mechanisms, or disease processes. To address the challenges posed by high dimensionality and multicollinearity, penalized variants incorporating L1 or L2 regularization are widely employed. These penalties shrink less informative coefficients toward zero, thereby stabilizing model estimates and performing implicit feature selection, which is especially valuable in SNP- or gene-level analyses [6, 7].

3.1.2 Margin-Based Classification for High-Dimensional Molecular Profiles

Support Vector Machines aim to identify a decision boundary that maximally separates classes in feature space, yielding robust classifiers even in high-dimensional settings. Using kernel functions, SVMs can model nonlinear relationships by implicitly mapping genomic features into higher-dimensional spaces. In genomics, SVMs have been applied to tasks such as classifying gene expression profiles across biological conditions, including disease versus control states. Their reliance on margin maximization makes them particularly effective when sample sizes are modest, but class separation is governed by complex multivariate patterns [6].

3.1.3 Ensemble Decision Frameworks for Genomic Feature Interaction Analysis

Decision trees classify samples by recursively partitioning the feature space using threshold-based rules, producing models that are intuitively interpretable as decision paths. However, individual trees are often unstable when applied to noisy or high-dimensional genomic data. Ensemble methods such as random forests mitigate this limitation by aggregating predictions from many trees trained on randomized subsets of samples and features. This aggregation strategy enhances predictive robustness and enables the modeling of nonlinear interactions among genomic variables. Importantly, tree-based ensembles can compute feature importance measures, which offer biologically meaningful insights by highlighting candidate genes, variants, or molecular signals associated with the outcome of interest [6, 7].

3.1.4 Instance-Based Similarity Methods in Genomic Classification

The k-nearest neighbors (kNN) algorithm is an instance-based method that assigns class labels according to the similarity between a query sample and its nearest neighbors in feature space. Although conceptually simple, kNN can be effective in genomic applications following dimensionality reduction or careful feature engineering. Nonetheless, its reliance on distance metrics makes it sensitive to the curse of dimensionality, necessitating preprocessing steps to ensure meaningful neighborhood structure in high-dimensional genomic spaces [6].

3.1.5 Probabilistic Inference Methods for Genomic Data Interpretation

Naive Bayes classifiers apply Bayes' theorem under the simplifying assumption that features are conditionally independent given the class label. While this assumption is often violated in genomic data, where linkage disequilibrium and co-expression introduce strong dependencies, Naive Bayes models can still perform competitively in practice. Their computational efficiency and probabilistic outputs make them useful as rapid baseline models or as components within ensemble frameworks, particularly during exploratory phases of genomic analysis [6].

Collectively, these classical ML models embody a trade-off between simplicity, interpretability, and predictive capacity. Their explicit optimization objectives such as likelihood maximization in logistic regression or margin maximization in SVMs yield models whose decisions can often be rationalized in terms of underlying biological signals rather than opaque latent representations.

3.1.6 Genomic Applications of Classical Machine Learning

This section shows how classical machine learning algorithms are used for various applications related to genomics.

- **Disease Risk Prediction and Variant Classification**: Classical ML methods have been applied to genotype and gene expression data to estimate disease risk, stratify patients, or classify molecular subtypes. By learning associations between genetic variants and phenotypic outcomes, these models support the prioritization of candidate variants for further biological investigation or functional validation.
- **Genomic Selection and Breeding Applications**: In agricultural genomics, regularized regression models and ensemble methods are widely used for genomic prediction of complex traits. These approaches enable the estimation of breeding values directly from marker profiles, reducing reliance on exhaustive phenotyping and accelerating selection cycles in plant and animal breeding programs [6].
- **Feature Selection and Biological Interpretation Pipelines**: Feature importance metrics derived from tree-based ensembles or coefficient magnitudes from penalized linear

models are frequently used to guide biological interpretation. Identifying genomic features that strongly influence model predictions can reveal regulatory regions, genes, or pathways worthy of targeted experimental follow-up.

- **RNA-Seq and Expression Pattern Classification**: Support Vector Machines and Random Forests have been successfully applied to RNA-seq datasets for classifying tissue types, disease states, or experimental conditions. These models assist in detecting expression patterns and molecular signatures associated with specific biological phenomena, complementing differential expression analyses [6, 7].

3.1.7 Strengths of Classical Machine Learning in Genomic Analysis

The strengths of classical machine learning in genomic analysis are as follows:

- **Interpretability**: Models such as logistic regression and decision trees provide transparent mechanisms for relating genomic features to predicted outcomes, facilitating biological insight.
- **Computational Efficiency**: Classical ML methods typically train and infer efficiently on standard computational resources, making them accessible for large-scale exploration studies.
- **Baseline Performance**: When combined with appropriate feature engineering, these models often achieve competitive performance and serve as reliable baselines for evaluating more complex AI approaches [6, 7].

3.1.8 Limitations and Motivations for Advanced AI Paradigms

The limitations and motivations for advanced AI paradigms are as follows:

- **Limited Representation of Complex Interactions**: Capturing higher-order or long-range genomic interactions is challenging without extensive kernelization or ensemble complexity.
- **Scalability Constraints**: Extremely high dimensionality can impair model stability and performance unless dimensionality reduction or regularization is carefully applied.
- **Dependence on Feature Engineering**: Model success is closely tied to expert-driven preprocessing, encoding strategies, and assumptions about feature relevance and independence.

While classical machine learning methods remain indispensable for genomic analysis due to their interpretability and efficiency, their limited capacity to model hierarchical and latent structure motivates the adoption of deep learning and transformer-based architecture. These modern approaches are built upon classical foundations by learning richer representations directly from raw genomic inputs and by exploiting complex dependencies across multi-omics data modalities. The following subsection examines how deep learning architectures address these challenges.

3.2 Deep Learning Architectures for DNA and RNA

Deep learning has emerged as a transformative paradigm for analyzing DNA and RNA data, enabling models to automatically learn hierarchical representations from raw biological sequences. Unlike classical machine learning approaches that rely on handcrafted features, deep learning architectures operate directly on nucleotide sequences, signal tracks, or expression profiles, discovering intricate patterns that are often inaccessible to manual feature engineering. This capability is particularly important in genomics, where regulatory logic is distributed across long genomic regions and influenced by local sequence context, epigenetic signals, three-dimensional chromatin organization, and cellular state [8, 9].

From a biological perspective, DNA and RNA sequences encode information across multiple scales. Short motifs dictate transcription factor binding and local RNA structural elements, medium-range interactions modulate splicing and RNA folding, and long-range dependencies influence gene expression through chromatin looping and enhancer-promoter interactions. Deep learning models are especially suited to capture these layered dependencies by stacking multiple nonlinear transformations, enabling abstraction from raw nucleotides to biologically meaningful representations [10]. By integrating sequence context at different scales, these models can reveal combinatorial and epistatic effects that are otherwise difficult to capture using classical approaches.

In practice, deep learning has shifted genomic analysis from rule-based modeling toward data-driven inference, allowing researchers to predict functional elements, regulatory activity, and transcriptomic outcomes directly from sequence information. This paradigm shift has accelerated discoveries in regulatory genomics, transcriptomics, and functional annotation while also raising new challenges in interpretability, generalization, and model robustness [11].

3.2.1 Model Architecture

Deep learning architectures for DNA and RNA analysis typically begin with an encoding stage, where nucleotide sequences are transformed into numerical representations. One-hot encoding remains common due to its simplicity and clear correspondence to biological sequences, while embedding-based representations are increasingly employed to capture contextual similarity among nucleotides or k-mers [12]. These embeddings allow the model to leverage subtle correlations and positional dependencies that may be critical for regulatory logic. At the core of these architectures are stacked neural layers that progressively learn higher-order abstractions. Convolutional layers are commonly used in early stages to detect local sequence patterns, analogous to biological motifs such as transcription factor binding sites. These layers function as trainable filters scanning sequences for recurring nucleotide combinations associated with regulatory or structural functions. Deeper layers integrate these local signals into global representations, enabling predictions of functional outcomes from distributed sequence features. For RNA-focused tasks, architectures are often adapted to account for variable sequence lengths, secondary structure, and temporal transcription dynamics. Models may integrate auxiliary inputs such as coverage profiles, splice junction signals, or epigenomic tracks to enrich representations. Multi-branch architecture facilitates the integration of heterogeneous modalities within a unified framework, enabling the simultaneous modeling of sequence, expression, and regulatory signals [13]. Training these architectures typically relies on supervised learning using labeled genomic datasets, such as experimentally validated binding sites, RNA splicing outcomes, or expression measurements. However, self-supervised and weakly supervised strategies are increasingly adopted to leverage vast repositories of unlabeled sequences, improving representation quality and enabling generalization across species, tissue types, or cellular contexts [14]. These strategies allow models to learn underlying sequence grammar and regulatory principles, even in the absence of explicit annotations.

3.2.2 Beyond CNNs: Hybrid and Alternative Architectures

While convolutional neural networks (CNNs) excel at local pattern recognition, additional architectures are often needed to capture long-range dependencies and complex epistatic relationships. Recurrent networks, transformers, state-space models, and specialized units such as Kolmogorov–Arnold networks (KANs) increasingly complement CNNs. Hybrid architectures such as CNN + BiLSTM, CNN + attention, or CNN + Mamba balance local motif detection with long-range context modeling, improving predictive performance while maintaining manageable complexity [14].

Emerging architectures, including Lyra and selective state-space models, offer subquadratic scaling for sequence length and strong performance across multiple genomic tasks. These designs highlight pathways toward efficient long-context genomic modeling, enabling regulatory prediction across megabase-scale regions without prohibitive computational cost. Table 3.2 shows different deep learning architectures and their genomic roles.

Table 3.2 Deep learning architectures and their genomic roles

Architecture family	Strengths in genomics	Typical use cases	Citations
CNNs	Local motif detection, computational efficiency	TF/RBP binding, chromatin marks, short-range regulatory logic	[8, 9]
CNN-hybrids (CNN + RNN/ attention/SSM)	Integration of local and long-range dependencies	Enhancer-promoter links, gene expression, variant effects	[12, 13]
Transformers/ gLMs	Contextualized embeddings, foundation model capabilities	Multi-task regulatory prediction, cross-species or region transfer	[14, 15]
Specialized (KANs, hyperbolic CNNs)	Expressive and parameter-efficient representations	Benchmarking, genome interpretation, TFBS prediction	[14, 15]

3.2.3 Genomic Applications

Deep learning architectures have been applied widely across DNA- and RNA-centric tasks. In regulatory genomics, models predict promoter and enhancer activity directly from DNA sequences, enabling genome-wide identification of functional regulatory elements [9, 15]. These predictions also facilitate variant interpretation by estimating the potential impact of sequence changes on regulatory logic.

In transcriptomics, deep models analyze RNA splicing, isoform abundance, and RNA stability. By capturing dependencies between sequence context and splicing outcomes, these models provide insights into alternative splicing mechanisms and their dysregulation in disease [10]. Similarly, models trained on RNA-seq data can predict expression levels from genomic sequence, linking genotype to phenotype at scale.

Deep architecture is also instrumental in functional annotation of non-coding regions, which comprise most of the human genome. By learning from experimental genomic datasets, these models infer functional relevance even in regions lacking direct experimental characterization. In applied biotechnology, such approaches support target identification, biomarker discovery, and prioritization of variants for experimental validation [14].

3.2.4 Strengths and Limitations

The primary strength of deep learning in DNA and RNA analysis lies in its ability to model complex, nonlinear relationships without explicit feature engineering. This flexibility enables the detection of subtle regulatory signals and long-range interactions that are challenging to encode manually. Furthermore, deep architectures scale effectively with increasing data volume, benefiting from the rapid growth of sequencing datasets [14, 15].

However, these advantages come with important limitations. Deep learning models generally require large, well-curated training datasets, which may be unavailable for rare conditions or underrepresented populations. Model interpretability remains a critical challenge, as learned representations are often opaque, complicating biological interpretation and translational utility [11]. Moreover, deep architectures may inadvertently learn dataset-specific biases, reducing robustness across experimental platforms, populations, or species. Mitigating these risks requires careful validation, rigorous benchmarking, and integration of domain knowledge into model design and evaluation [15].

While general-purpose deep learning architecture provides a powerful foundation for genomic analysis, many DNA- and RNA-specific tasks rely heavily on detecting localized sequence patterns, such as motifs and regulatory elements. Convolutional neural networks (CNNs) have become a cornerstone for sequence pattern recognition in genomics. The next subsection focuses on CNN-based architectures, exploring how convolutional layers are tailored to extract biologically meaningful patterns from genomic sequences at scale.

3.3 Convolutional Neural Networks for Sequence Patterns

Convolutional Neural Networks (CNNs) have emerged as a foundational architecture for extracting hierarchical representations from structured genomic data. Originally developed for image processing, CNNs leverage local connectivity and weight sharing to detect patterns that are invariant to translation along the input. In genomics, DNA and RNA sequences can be interpreted as one-dimensional arrays of nucleotides, where short subsequences frequently encode functional elements such as motifs, transcription factor binding sites, splice signals, or regulatory motifs. CNNs identify these patterns by convolving learnable filters across the input sequence, effectively learning sequence features directly from data rather than relying on predefined scoring schemes or motif libraries [16, 17]. CNNs impose an inductive bias appropriate for biological sequences, assuming that functional patterns are local and compositional. Short sequence motifs often combine recurrently to determine higher-order regulatory outcomes, such as gene expression, chromatin accessibility, or splicing efficiency. By stacking convolutional layers with pooling and nonlinear transformations, CNNs progressively abstract simple motifs into complex hierarchical features. This property enables end-to-end learning of predictive sequence representations that generalize across genomic contexts without manual feature engineering [16, 17].

3.3.1 Architectural Principles and Mechanisms

A typical CNN for genomic sequences consists of the following components:

- **Input Layer**: One-hot encoded nucleotides (A, C, G, T) or fixed-length embeddings of k-mers.
- **Convolutional Layers**: Apply filters along the sequence to detect recurring patterns predictive of a biological outcome. Early layers capture low-level motifs, while deeper layers combine motifs into higher-order regulatory signatures.
- **Activation Functions**: Nonlinearities such as ReLU enable modeling of complex relationships beyond simple motif presence.
- **Pooling Layers**: Summarize features over broader sequence regions, introduce positional invariance, and reduce computational load.
- **Output Layer**: Fully connected or task-specific heads (e.g., classification, regression) map learned representations to predictions, such as binding site occupancy or regulatory function.

Modern variants incorporate dilated convolutions and gated units to expand the receptive field, capturing longer-range dependencies without excessively deep networks. Hybrid architecture combining CNNs with recurrent layers (e.g., LSTMs) or attention mechanisms address distal interactions, reflecting the combinatorial logic of genomic regulation [18]. Table 3.3 shows the various deep learning models and how these models are used in genomic-related applications.

This comparison situates CNNs within the broader deep learning toolkit, emphasizing that while CNNs excel at local sequence pattern detection, their limitations in modeling long-range interactions motivate complementary architectures, such as recurrent or transformer-based networks. These architectures are explored in Sect. 3.4, bridging local motif recognition with global functional inference.

3.3.2 Genomic Applications

CNNs have been applied extensively across DNA- and RNA-centric tasks where sequence pattern detection is crucial:

- **Transcription Factor Binding Sites (TFBS) Prediction**: CNNs learn binding motifs and contextual dependencies directly from sequences, outperforming traditional motif scoring methods. Advanced architectures integrate multiple CNN experts and interpret model decisions via attribution mapping [19].

Table 3.3 Deep learning models and their genomic roles

Model class	Representative models	Primary strengths	Primary limitations	Recommended use cases
Pure CNNs	Basset, DeepSEA, TREDNet, SEI	Local pattern recognition, computational efficiency, robust across cell types	Limited receptive field, cannot capture very long-range interactions	TF binding prediction, chromatin accessibility
Hybrid CNN-RNN	DanQ	Captures motif dependencies, regulatory grammar	Computationally intensive, longer training	Noncoding function prediction, regulatory element classification
Hybrid CNN-Transformer	Enformer, Borzoi	Long-range interactions, enhancer-promoter prediction, causal SNP identification	High computational cost may require broader training	Gene expression prediction, variant effect prediction, fine-mapping GWAS loci
Transformers/ Foundation Models	Nucleotide Transformer, DNABERT series	Transfer learning, effective in low-data regimes, unsupervised regulatory discovery	Fine tuning is often required, limited fine gained signal resolution	Cross species applications, variant prioritization, low-data regulatory tasks

- **Gene Expression Prediction**: By relating raw DNA sequences to expression levels, CNNs reveal regulatory grammar in non-coding regions, contributing to mechanistic understanding of transcriptional regulation [20].
- **Sequence Classification and Variant Profiling**: CNNs classify sequences into functional categories (promoters, enhancers, viral strains) and identify discriminative motifs or mutational hotspots. Visualization techniques, such as saliency maps, provide biologically interpretable insights [21].
- **Epigenomic Mark Prediction**: Local sequence preferences underlying histone modifications, DNA methylation, and chromatin accessibility are detected by CNN filters, informing epigenomic models.
- **Motif Discovery and Regulatory Annotation**: CNNs facilitate de novo motif identification and functional annotation, converting learned filters into sequence logos for biological interpretation. This capability aids genome annotation in poorly characterized organisms or cell types.

Across these applications, CNNs serve both predictive and hypothesis-generating roles, enabling discovery of novel regulatory patterns for experimental follow-up.

3.3.3 Strengths and Limitations

The strengths and limitations of deep learning are as follows:

- **Strengths**:
  - **Hierarchical Feature Extraction**: Aligns with the modular structure of genomic regulatory code.
  - **Efficient Computation**: Weight sharing and sparse connectivity mitigate overfitting in high-dimensional sequences.
  - **Interpretability**: Early-layer filters correspond to motifs; attribution techniques highlight influential subsequences.
  - **End-to-End Learning**: Reduces dependence on handcrafted features, supporting exploratory genomics.

- **Limitations**:
  - **Receptive Field Constraints**: Standard CNNs may miss long-range dependencies spanning kilobases of sequence. Hybrid CNN-RNN or CNN-attention models address this at the cost of increased complexity [18].
  - **Abstract Feature Interpretation**: Deeper layers capture complex patterns that are less directly interpretable.
  - **Data Dependence**: Model performance is sensitive to the quality, balance, and representativeness of training datasets. Biases in well-characterized regions can limit generalization.

3.4 Recurrent and Attention-Based Models

Recurrent and attention-based architectures occupy a central role in modern AI-driven genomic analysis. These methods are designed to capture sequential and contextual information inherent in DNA, RNA, and multi-omics datasets, enabling tasks such as functional annotation, regulatory motif discovery, variant prioritization, and sequence classification. Unlike CNNs, which excel at local pattern recognition, recurrent and attention-based models explicitly model long-range dependencies and contextual relationships, reflecting the distributed regulatory logic of the genome [22, 23].

3.4.1 Conceptual Overview

Genomic sequences are intrinsically sequential: the biological function of a nucleotide or regulatory element often depends on distant signals such as enhancers, silencers, or chromatin interactions. Classical and even standard deep learning approaches that rely on local motifs may fail to capture these distal dependencies comprehensively [22, 23].

Recurrent Neural Networks (RNNs) were among the first architectures adapted for sequential genomics. They maintain a hidden state that evolves as nucleotides are processed step by step, enabling the network to retain information from earlier positions. LSTM and GRU variants mitigate challenges such as vanishing gradients, allowing the network to learn longer-range dependencies while maintaining biologically meaningful representations of distal regulatory signals [22, 23].

Attention-based models, exemplified by Transformers, represent a conceptual shift. Instead of processing sequences sequentially, self-attention layers compute relationships among all positions simultaneously. This enables the model to "attend" selectively to relevant nucleotides or motifs, regardless of distance along the sequence. Multi-head attention extends this idea by capturing diverse interactions, supporting complex regulatory inference [23, 24]. Positional encodings or learned embeddings reintroduce sequence order information, ensuring that the model retains biologically meaningful spatial context [24].

Biological Intuition: Self-attention allows models to highlight regulatory motifs or structural elements that may be separated by tens of kilobases, reflecting distal enhancer-promoter interactions, long-range chromatin loops, or overlapping regulatory modules. RNNs, with gating mechanisms, offer a complementary approach by sequentially propagating context along the genome [22, 23].

3.4.2 Architectural Principles

- **Recurrent Architectures**:
  - Iteratively update a hidden state as each nucleotide is processed.
  - Standard RNNs capture short- to medium-range dependencies; LSTM/GRU units enable longer-range information retention.
  - Useful for scanning along sequences where the influence of upstream elements persists over several kilobases.

- **Attention-Based/Transformer Architectures**:
  - Compute weighted contributions of all positions in a sequence simultaneously.
  - Multi-head attention allows parallel modeling of multiple dependencies or motifs.
  - Positional encodings preserve nucleotide order while enabling highly parallelized computation.
  - Facilitate long-range interaction modeling, essential for regulatory genomics.

Hybrid architecture combining RNNs with attention layers exploits complementary strengths: sequential context propagation and global dependency capture [25]. Bidirectional training paradigms further reflect the dual-direction nature of DNA, capturing interactions across both strands [24].

3.4.3 Genomic Applications

- **Sequence Classification and Variant Prioritization**:
  - Classify genomic regions, promoters, enhancers, or viral sequences.
  - Attention weights highlight biologically important positions, supporting prioritization of pathogenic variants.
  - Hybrid attention-RNN models have demonstrated strong performance in pathogen genomics, including SARS-CoV-2 variant classification [25].

- **Regulatory Sequence Modeling**:
  - Predict gene regulatory outputs from raw DNA or RNA sequence.
  - Foundation models pretrained on large genomic datasets integrate self-attention to capture distal regulatory grammar at single-nucleotide resolution [23, 24].
  - Bidirectional architectures model overlapping genes and bidirectional promoters, enhancing biological plausibility [24].

- **Multi-omics and Integrative Analysis**:
  - Attention mechanisms enable simultaneous integration of DNA sequence, histone marks, methylation profiles, and gene expression.
  - Facilitates phenotype prediction and cellular response modeling in complex regulatory landscapes, supporting multi-omic studies of disease mechanisms [5, 26, 27].

3.4.4 Strengths and Limitations

The strengths and limitations of these models are as follows:

- **Strengths**:
  - **Long-Range Context Capture**: Attention mechanisms excel at modeling distal interactions spanning kilobases to megabases [23, 24].
  - **Parallel Training Efficiency**: Transformers allow large-scale genomic pretraining with highly parallel computation [24].
  - **Interpretability**: Attention scores and RNN hidden states can be interrogated to identify influential sequence positions, providing mechanistic insights [22, 23].

- **Limitations**:
  - **Computational Complexity**: Self-attention scales quadratically with sequence length, complicating full-chromosome analysis without approximations [23].
  - **Data Requirements**: Large pretrained models require extensive genomic datasets and high-performance computing [24, 26].
  - **Biological Biases**: Models trained on reference genomes may embed population-specific biases, limiting cross-species generalization [26].
  - **RNN Constraints**: Despite gating mechanisms, RNNs struggle with very long dependencies and require extensive training for complex genomic sequences [22, 23].

Recurrent and attention-based models extend the capabilities of CNNs by capturing sequential and long-range dependencies, supporting integrative and context-aware genomic analyses. They form a natural foundation for generative and multi-modal models, which are explored in the next section, where synthetic sequence generation and latent representation learning further complement predictive genomics.

3.5 Transformers and Foundation Models for Biology

The advent of transformer architectures and large-scale foundation models has transformed the computational study of genomic sequences. Initially developed for natural language processing (NLP), transformers treat nucleotides or amino acid residues as tokens within a sequence, enabling models to learn complex contextual dependencies analogous to linguistic syntax and semantics [28, 29]. This analogy allows the capture of long-range regulatory interactions, such as enhancer–promoter communication or distal splicing signals, which are often inaccessible to classical machine learning or convolutional approaches [29, 30].

Foundation models extend transformers by pretraining on massive unlabeled genomic datasets using self-supervised objectives, producing generalizable sequence representations. These pretrained embeddings can subsequently be fine-tuned for diverse downstream tasks including regulatory element prediction, variant effect assessment, or multi-omics integration often with limited labeled data [23, 31].

3.5.1 Transformer Model Architecture

Transformers comprise self-attention layers, positional encodings, and feedforward networks. Self-attention enables context-sensitive embeddings by calculating pairwise interactions between all sequence positions, essential for capturing distal regulatory dependencies spanning tens of thousands of base pairs [32]. Positional encodings preserve sequence order, ensuring that the model retains biologically meaningful context, while

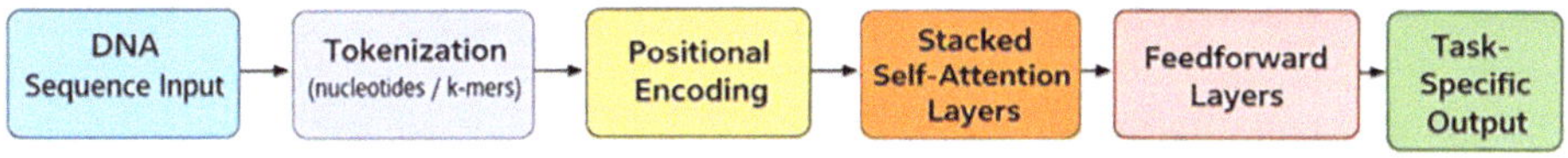

Fig. 3.1 Schematic of a transformer architecture applied to genomic sequences

feedforward layers transform these representations into higher-level features suitable for downstream tasks [33] (Fig. 3.1).

- **Self-attention and Contextual Embeddings**: Self-attention evaluates similarity scores across all token pairs and uses these weights to generate contextualized embeddings. This capability allows transformers to model complex regulatory grammar that cannot be captured by k-mer statistics or motif scanning alone, including chromatin-mediated interactions and co-occurring transcription factor motifs [33]. Multi-head attention further enables the simultaneous capture of multiple functional patterns. Table 3.4 shows the pretraining paradigm of different models.
- **Scaling and Tokenization**: Transformer scalability through larger parameter counts, extended sequence contexts, and k-mer tokenization directly improves the representational richness of genomic embeddings. Simple single-nucleotide tokens retain maximal resolution, whereas variable-length k-mers capture recurring motifs or codons efficiently. Larger context windows allow modeling of long-range chromatin interactions and evolutionary conservation patterns across species [35].

Table 3.4 Pertaining paradigms

Pretraining paradigm	Mechanism	Genomic intuition
Masked Language Modeling (MLM)	Randomly masks segments of the sequence and predicts them from surrounding context	Encourages bidirectional understanding of sequence dependencies (e.g., flanking motifs or secondary RNA structure) [34]
Causal Language Modeling	Autoregressively predicts the next nucleotide	Supports sequence generation tasks, simulating mutational or transcriptional processes [34]
Hybrid/Bidirectional + Autoregressive	Combines MLM and autoregressive objectives	Captures both predictive and generative properties of sequences, reflecting biological directionality (e.g., transcription or RNA splicing) [24]

3.5.2 Genomic Use Cases

The various applications of transformer models in genomics are as follows (shown in Fig. 3.2):

- **Regulatory Element and Functional Annotation**: Pretrained embeddings capture conserved sequence and regulatory patterns, enabling fine-tuning for detection of promoters, enhancers, and splice junctions. These embeddings outperform classical motif scanning or CNN-based sequence classifiers by integrating both local motifs and global context [36].
- **Variant Effect Prediction**: Foundation models improve the interpretation of non-coding variants by evaluating subtle changes in sequence context, capturing effects that may only manifest through distal interactions [37].
- **Multi-modal and Cross-task Integration**: Transformers facilitate integration of DNA sequence, epigenomic signals, and transcriptional profiles, producing holistic embeddings for systems-level understanding of gene regulation [38].
- **RNA Regulatory Landscape Analysis**: Language model paradigms extend naturally to RNA, where nucleotide sequence determinants of splicing, translation, and degradation are encoded in context-dependent embeddings. This enables modeling of post-transcriptional regulatory grammar, crucial for disease genomics and functional studies [39].

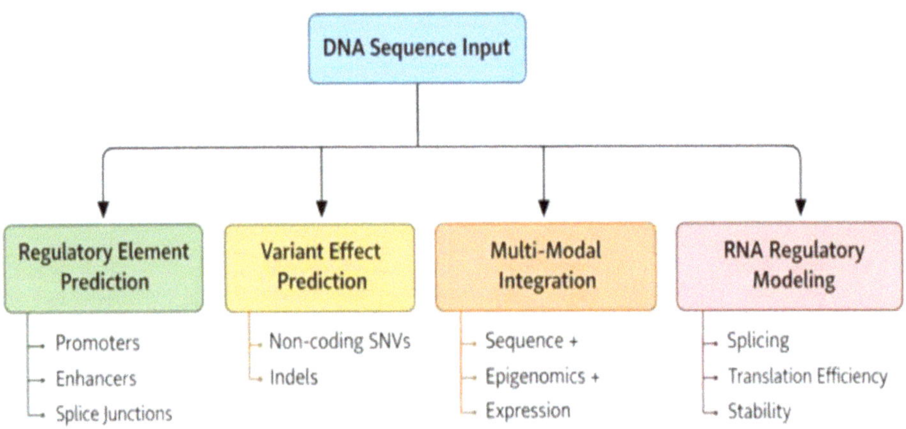

Fig. 3.2 Transformers and foundation model applications in genomics

3.5.3 Strengths and Limitations

The strengths and limitations of these models are given in Tables 3.5 and 3.6.

Transformers and foundation models represent a paradigm shift in computational genomics, complementing CNNs (3.3) and attention-based models (3.4) by scaling self-supervised learning to massive datasets and enabling cross-task, multi-modal embeddings. They serve as a bridge between pattern-focused models and data-driven, integrative genomic representations, setting the stage for generative modeling, synthetic sequence design, and the integration of knowledge graphs and pathway information, which will be explored in Sect. 3.6.

Table 3.5 Key strengths of transformer-based models for biological sequence analysis

Strengths	Biological relevance	References
Generalization and Transferability	Pretrained embeddings capture generic regulatory logic, reducing the need for large, labeled datasets	[40]
Long-range dependency learning	Self-attention models distal interactions, enabling enhancer-promoter and chromatin architecture modeling	[32]
Multi-task and Multi-modal Integration	Jointly models sequence, epigenomics, and transcriptomics	[38]

Table 3.6 Key limitations of transformer-based models for biological sequence analysis

Limitations	Impact on genomic modeling	References
Computational requirements	Large models require high-performance computing; smaller labs may face resource constraints	[28]
Interpretability	Large-scale embeddings are opaque; attribution of predictions to biological motifs is challenging	[41]
Bias and domain generalization	Models trained on limited reference genomes may not generalize across populations or species	[40]
Evaluation pitfalls	Differences in fine-tuning, pooling, and layer selection can bias benchmarking	[41]

3.6 Generative AI for Sequence Design

Generative artificial intelligence (AI) has emerged as a transformative paradigm in genomic analysis, shifting the focus from predicting properties of existing sequences to designing novel sequences with desired biological functions. Unlike predictive models, which evaluate the function of known sequences, generative models aim to create de novo DNA, RNA, or protein sequences that meet functional objectives. These approaches repurpose deep learning architectures originally developed for language, vision, or image generation to learn the complex statistical distributions inherent in biological sequences, enabling the sampling of biologically plausible candidates [43, 44].

Biological sequences can be conceptualized as structured "languages," with nucleotides or amino acids serving as discrete tokens governed by syntactic and semantic rules. Generative AI captures local motifs, such as transcription factor binding sites, and long-range dependencies, including distal promoter–enhancer interactions, splicing signals, or RNA secondary structures. By learning these patterns from large genomic datasets, models can explore previously uncharted regions of sequence space while maintaining biological plausibility [43, 44].

This approach is particularly valuable given the astronomical size of sequence space: even modest-length proteins or regulatory elements have combinatorial complexity beyond exhaustive experimental screening. Generative models provide a computational solution, proposing sequences that balance diversity, functional potential, and adherence to natural constraints, accelerating research in synthetic biology, therapeutic development, and functional genomics [46].

3.6.1 Model Architecture and Mechanisms

Modern generative sequence design predominantly relies on transformer-based architectures, variational autoencoders (VAEs), and diffusion models (example shown in Fig. 3.3):

- **Transformers**: Encode dependencies across tens of thousands of nucleotides, capturing both local and distal interactions. Conditional transformers allow context-guided

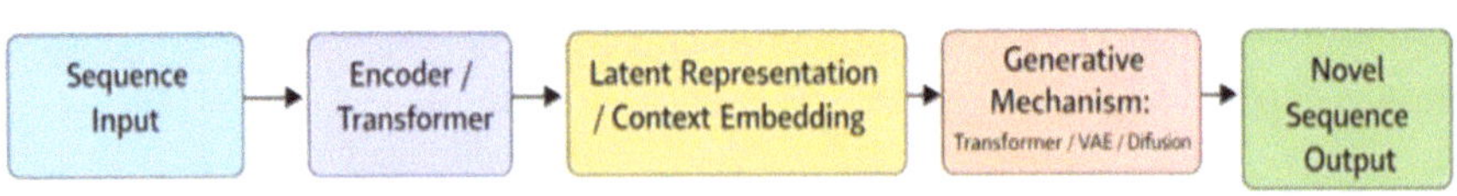

Fig. 3.3 Generative AI Architectures for Sequence Design

sequence generation, like the attention mechanisms discussed in 3.4 and foundation models in 3.5 [44].

- **Variational Autoencoders (VAEs)**: Compress sequences into latent representations, enabling interpolation between functional sequences and generation of novel candidates by decoding latent vectors [45].
- **Diffusion Models**: Iteratively denoise random sequences toward biologically realistic configurations, preserving structural and functional constraints while offering high sequence diversity [48].

3.6.2 Genomic Use Cases

The applications of generative AI in genomics are as follows (as shown in Fig. 3.4):

- **De Novo Regulatory Element Design**: Generative models can produce synthetic DNA regulatory elements, including promoters and enhancers, with controlled expression profiles. By learning patterns from natural regulatory sequences and associated functional outputs, models can propose sequences achieving desired gene expression levels, cell-type specificity, or context-dependent regulation [42].
- **RNA Sequence Engineering**: RNA molecules carry out structural, regulatory, and catalytic functions. Generative AI can design RNA sequences predicted to fold into defined secondary or tertiary structures or bind specified protein partners. Applications include RNA aptamer design, regulatory RNA optimization, and therapeutic constructs [43].
- **Protein Sequence Design**: Generative models trained on proteomic datasets generate novel protein sequences with predicted structural stability or functional activity. These

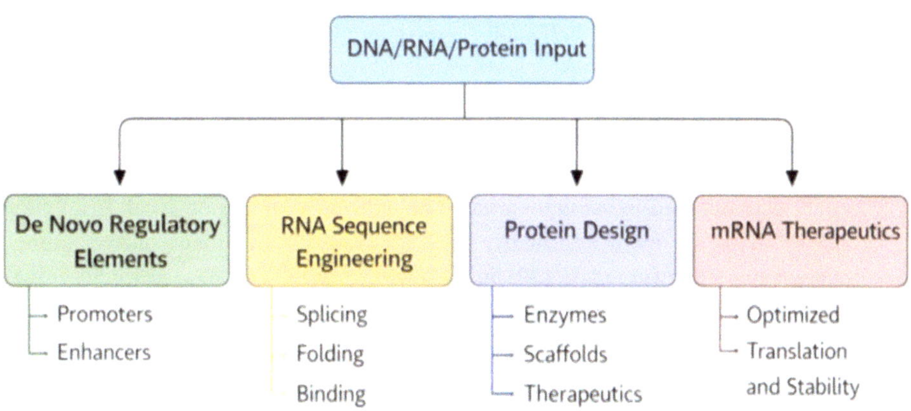

Fig. 3.4 Overview of generative AI applications in sequence design

models explore previously untested regions of protein space, offering candidates for enzymes, binding proteins, or scaffolds in synthetic biology [47].
- **mRNA Therapeutic Optimization**: In mRNA-based therapeutics, generative AI helps optimize sequences for translational efficiency, stability, and codon usage, guided by conditional generation or reinforcement learning. This supports the design of functional mRNA constructs that are biologically robust [43].

3.7 Model Evaluation: Metrics, Validation, and Bias

Evaluating artificial intelligence (AI) models in genomics is essential for ensuring predictions are robust, biologically meaningful, and generalizable. AI applications range from variant effect prediction, disease risk stratification, and cell-type classification to sequence annotation and regulatory element prediction. Each task demands careful selection of metrics, validation strategies, and bias analyses, reflecting the complexity and heterogeneity of genomic data. Rigorous evaluation safeguards against overfitting, data leakage, and unaccounted biases, ensuring that models provide actionable biological insights [49, 50, 52]. Model evaluation in genomics encompasses three interrelated dimensions:

1. Performance metrics, quantifying model quality and predictive power.
2. Validation strategies, assessing generalization to unseen data.
3. Bias assessment, identifying systematic disparities in model performance across populations, sequencing platforms, or biological contexts.

3.7.1 Performance Metrics in Genomic AI

- **Classification and Ranking**:
  - Many genomic AI tasks, such as distinguishing pathogenic from benign variants or predicting disease phenotypes, are formulated as classification problems. Standard metrics include Accuracy, Precision, Recall, F1-score, ROC-AUC (Area under the Receiver Operating Characteristic curve).
  - Class imbalance is common in genomics; for example, rare pathogenic variants are vastly outnumbered by benign variants. In such cases, precision-recall (PR) curves and PR-AUC are more informative than accuracy, as they focus on the minority, biologically significant class [49, 50].

- **Regression and Continuous Outcomes**:
  - Genomic models may also predict continuous traits, such as gene expression levels, quantitative traits, or epigenetic scores. Relevant metrics include Mean Squared Error (MSE), Mean Absolute Error (MAE), Explained Variance (R^2).

  - These metrics quantify the closeness of predicted and observed values but must be interpreted considering biological noise and measurement variability, which can limit maximum achievable accuracy [49].
- **Clustering and Unsupervised Measures**:
  For unsupervised analyses (e.g., clustering of single-cell RNA-seq data), evaluation differs:
  - **Intrinsic Metrics**: Assess cluster compactness and separation without labels.
  - **Extrinsic Metrics**: Compare predicted clusters with known biological annotations.
  - **Caution**: Numeric clustering scores do not guarantee biological relevance; expert knowledge must guide interpretation [52].
  - **Calibration and Decision Support**: For probabilistic models (e.g., disease risk prediction), calibration ensures predicted probabilities match observed outcomes. Key tools include Calibration curves and Brier scores Well-calibrated outputs are critical when AI models inform clinical or experimental decisions [52].

3.7.2 Validation Strategies for Robust Assessment

Metrics alone are insufficient without robust validation, as genomic datasets often feature linkage disequilibrium, population stratification, or familial relationships that can bias performance estimates.

- **Cross-Validation and Hold-Out Testing**:
  - K-fold cross-validation estimates performance reliably for limited sample sizes.
  - Stratified cross-validation ensures proportional representation of key subgroups across folds.
  - Hold-out test sets, untouched during training, provide an unbiased assessment of generalization [49].

- **External and Independent Validation**: Testing on independent cohorts is essential. For example:
  - Models trained on European ancestry cohorts may underperform in populations with different allele frequencies.
  - External validation ensures predictions reflect biologically generalizable patterns rather than dataset-specific artifacts [50, 51].

- **Biological Benchmarking**: Evaluates whether predictions align with known biological mechanisms. For example:
  - Predicted regulatory elements confirmed with functional assays.
  - Variant effect predictions tested against independent experimental datasets [50].

Table 3.7 Key limitations of transformer-based models in genomic modeling

Source	Description	Implications
Representation bias	Underrepresentation of certain ancestries or phenotypes	Unequal performance, misinterpretation in diverse populations
Sampling bias	Cohort selection or recruitment practices	Overfitting to specific population features
Technical bias	Batch effects, sequencing platform differences, variant calling pipelines	Spurious patterns or inflated predictive confidence [52, 53]

3.7.3 Bias in Genomic AI: Identification and Implications

Bias occurs when models perform unevenly across populations, sexes, disease subtypes, or technical conditions. Table 3.7 shows key sources of bias and their implications in genomic modeling.

Representation bias is particularly critical in genomics due to population-level genetic diversity.

- **Assessing Bias**: Bias assessment involves the points below:

1. Stratifying performance metrics across subgroups (ancestry, sex, disease type).
2. Examining sensitivity, specificity, calibration, and false positive/negative rates.
3. Identifying miscalibration in probabilistic outputs, which can undermine trust in predictions [52].

3.7.4 Common Pitfalls and Best Practices

The common pitfalls and best practices used for model development and evaluation in the genomics are given in Table 3.8.

- **Best Practices**:

Table 3.8 Common pitfalls in genomic model development and evaluation

Pitfall	Description
Overfitting	Memorization of training noise leading to poor generalization
Data leakage	Accidental use of test information during model development
Single-metric reliance	Misleading evaluation if only one metric is reported

Table 3.9 Practical genomic applications and their evaluation strategies

Application	Evaluation approach	References
Newborn screening integrating genomic and metabolomic data	ROC-AUC, precision-recall, cross-validation, external cohort validation	[50]
Oncology variant interpretation	Precision-recall metrics for class imbalance, validation across external datasets	[51]
Transcriptomic classifiers	Precision-recall metrics for class imbalance, validation across external datasets	[53]

1. Report multiple complementary metrics (classification, calibration, PR-AUC).
2. Conduct rigorous cross-validation and independent testing.
3. Perform subgroup analyses to detect bias.
4. Ensure reproducibility with transparent reporting of datasets, preprocessing, and evaluation protocols [49, 52, 53].

3.7.5 Practical Examples in Genomics

Table 3.9 shows practical genomic applications and their evaluation strategies.

3.8 Summary

This chapter reviews AI methodologies in genomics, tracking progress from traditional statistical models to modern deep learning and generative frameworks. Genomic data presents analytical challenges such as high dimensionality and complex biological structure, demanding both robust computational strategies and careful interpretation.

Classical machine learning methods like logistic regression and Random Forests offer interpretable tools for classification and prediction but have limitations with nonlinear complexities. Deep learning, particularly CNNs and RNNs, improves feature extraction and sequential modeling in genomic sequences, although issues remain with long-range interactions and scalability.

Attention mechanisms and transformer-based architecture address these challenges by focusing on relevant sequence regions and conceptualizing DNA/RNA as structured languages, enhancing context-driven analysis. Foundation models enable self-supervised learning across broad tasks, expanding access but raising questions about computational demands and bias.

Generative AI facilitates the creation of novel DNA, RNA, and protein sequences, opening new avenues for synthetic biology, though experimental validation and ethical oversight are necessary. The chapter underscores the importance of thorough model evaluation, including task-appropriate metrics, cross-validation, and bias assessment, to ensure reliability and equity.

Interpretability is vital for trust, especially in clinical contexts, and should be integrated into model design. Future directions include scalable sequencing analysis, multi-omics integration, automated modeling, privacy-preserving techniques, and evolving ethical considerations. AI's growing influence in genomics brings scientific and societal opportunities and responsibilities.

References

1. Changalidis, A., Barbitoff, Y., Nasykhova, Y., & Glotov, A. (2026). A systematic review on the generative AI applications in human medical genetics. *Frontiers in Genetics, 16*, 1694070.
2. Thanki, R., & Shah, S. (2026). Artificial intelligence and its role in early detection of oral cancer. In *Artificial Intelligence for Early Detection and Diagnosis of Oral Cancer* (pp. 41–70). Springer Nature Switzerland.
3. Roberts, M. D., Davis, O., Josephs, E. B., & Williamson, R. J. (2025). K-mer-based approaches to bridging pangenomics and population genetics. *Molecular Biology and Evolution, 42*(3), msaf047.
4. Taddese, A. A., Addis, A. C., & Tam, B. T. (2025). Data stewardship and curation practices in AI-based genomics and automated microscopy image analysis for high-throughput screening studies: Promoting robust and ethical AI applications. *Human Genomics, 19*(1), 16.
5. Zhang, S., Fan, R., Liu, Y., Chen, S., Liu, Q., & Zeng, W. (2023). Applications of transformer-based language models in bioinformatics: a survey. *Bioinformatics Advances, 3*(1), vbad001.
6. Lu, J., Lu, X., Wang, Y., Zhang, H., Han, L., Zhu, B., & Wang, B. (2025). Comparison between logistic regression and machine learning algorithms on prediction of noise-induced hearing loss and investigation of SNP loci. *Scientific Reports, 15*(1), 15361.
7. Akter, S., Adesola, R. O., & Basnet, S. (2025). Machine learning approach to identify significant genes and classify cancer types from RNA-seq data. *Global Medical Genetics*, 100079.
8. Koumakis, L. (2020). Deep learning models in genomics; Are we there yet? *Computational and Structural Biotechnology Journal, 18*, 1466–1473.
9. Ismail, F. N., Sengupta, A., & Amarasoma, S. (2025). Deep learning for regulatory genomics: A survey of models, challenges, and applications. *Bioinformatics Advances*, vbaf271.
10. Apicco, D. J., Zhang, C., Maziuk, B., Jiang, L., Ballance, H. I., Boudeau, S., … & Wolozin, B. (2019). Dysregulation of RNA splicing in tauopathies. *Cell Reports, 29*(13), 4377–4388.
11. Munn, L. L., & Jain, R. K. (2025). Challenges and opportunities for the next generation of computational tumor models. *PLoS Biology, 23*(7), e3003269.
12. Iuchi, H., Matsutani, T., Yamada, K., Iwano, N., Sumi, S., Hosoda, S., … & Hamada, M. (2021). Representation learning applications in biological sequence analysis. *Computational and Structural Biotechnology Journal, 19*, 3198–3208.
13. Wang, Y., & Chen, L. (2022). DeepPerVar: A multi-modal deep learning framework for functional interpretation of genetic variants in personal genome. *Bioinformatics, 38*(24), 5340–5351.
14. Gündüz, H. A., Binder, M., To, X. Y., Mreches, R., Bischl, B., McHardy, A. C., … & Rezaei, M. (2023). A self-supervised deep learning method for data-efficient training in genomics. *Communications Biology, 6*(1), 928.

15. Ektefaie, Y., Shen, A., Bykova, D., Marin, M. G., Zitnik, M., & Farhat, M. (2024). Evaluating generalizability of artificial intelligence models for molecular datasets. *Nature Machine Intelligence, 6*(12), 1512–1524.
16. Bo, Y., Mao, W., Shao, Y., Bai, W., Ye, P., Ma, X., ... & Shen, C. (2025). Revisiting convolution architecture in the realm of DNA foundation models. arXiv preprint arXiv:2502.18538
17. Zhang, Z., Zhao, Y., Liao, X., Shi, W., Li, K., Zou, Q., & Peng, S. (2019). Deep learning in omics: A survey and guideline. *Briefings in Functional Genomics, 18*(1), 41–57.
18. Tabane, E., Mnkandla, E., & Wang, Z. (2025). Optimizing DNA sequence classification via a deep learning hybrid of LSTM and CNN architecture. *Applied Sciences, 15*(15), 8225.
19. Tripathi, A., Nielsen, I. E., Umer, M., Ramachandran, R. P., & Rasool, G. (2025). Explainable AI in Genomics: Transcription factor binding site prediction with mixture of experts. ArXiv, arXiv-2507.
20. Barbadilla-Martínez, L., Klaassen, N., van Steensel, B., & de Ridder, J. (2025). Predicting gene expression from DNA sequence using deep learning models. *Nature Reviews Genetics, 26*(10), 666–680.
21. Awe, O. I., Obura, H., Ssemuyiga, C., Mudibo, E., & Mwanga, M. J. (2025). Enhanced deep convolutional neural network for SARS-CoV-2 variants classification. *Frontiers in Artificial Intelligence, 8*, 1512003.
22. Vishniakov, K., Amor, B. B., Tekin, E., ElNaker, N. A., Viswanathan, K., Medvedev, A., ... & Pimentel, M. (2025). Gene42: long-range genomic foundation model with dense attention. arXiv preprint arXiv:2503.16565.
23. Ding, K., Parker, B., & Wen, J. (2025). NucEL: Single-nucleotide ELECTRA-style genomic pre-training for efficient and interpretable representations. arXiv preprint arXiv:2508.13191.
24. Duan, Q., Huang, B., Song, Z., Lehmann, I., Gu, L., Eils, R., & Wild, B. (2025). JanusDNA: a powerful bi-directional hybrid DNA foundation model. arXiv preprint arXiv:2505.17257.
25. Mutawa, A. M. (2025). Attention-based hybrid deep learning models for classifying COVID-19 genome sequences. *AI, 6*(1), 4.
26. de Almeida, B. P., Lopez, M., & Pierrot, T. (2025). AI models for genomics applications. *Nature Methods, 22*, 231–232.
27. Choi, S. R., & Lee, M. (2023). Transformer architecture and attention mechanisms in genome data analysis: A comprehensive review. *Biology, 12*(7), 1033.
28. Consens, M. E., Dufault, C., Wainberg, M., Forster, D., Karimzadeh, M., Goodarzi, H., ... & Wang, B. (2025). Transformers and genome language models. *Nature Machine Intelligence, 7*(3), 346–362.
29. Abdelwahab, O., & Torkamaneh, D. (2025). A Transformers-based framework for refinement of genetic variants. *Frontiers in Bioinformatics, 5*, 1694924.
30. Zhou, H., Hu, Y., Zheng, Y., Li, J., Peng, J., Hu, J., ... & Wang, Z. (2025). A foundation language model to decipher diverse regulation of RNAs. *Genome Biology, 26*(1), 301.
31. Li, Z., Subasri, V., Shen, Y., Li, D., Zhao, Y., Stan, G. B., & Shan, C. (2025). Omni-dna: A unified genomic foundation model for cross-modal and multi-task learning. arXiv preprint arXiv: 2502.03499.
32. Transfer Learning. https://en.wikipedia.org/wiki/Transformer_(deep_learning). Last Access: January 2026.
33. Balakrishnan, P., Anny Leema, A., Dhivya Shree, V., Mohammad Saad, C., & Mohan Babu, A. (2025). Gene-LLMs: A comprehensive survey of transformer-based genomic language models for regulatory and clinical genomics. *Frontiers in Genetics, 16*, 1634882.
34. DNA Large Language Model. https://en.wikipedia.org/wiki/DNA_large_language_model. Last Access: January 2026.

35. Dalla-Torre, H., Gonzalez, L., Mendoza-Revilla, J., Lopez Carranza, N., Grzywaczewski, A. H., Oteri, F., … & Pierrot, T. (2025). Nucleotide transformer: Building and evaluating robust foundation models for human genomics. *Nature Methods, 22*(2), 287–297.
36. Mousa, R., Sarabadani, A., Taami, T., Bengari, A. A., Eslamifar, O., Shalmani, M. A., & Shahmarvandi, E. K. (2025). *A comparative survey on large language models for biological data.*
37. Li, S., Fan, W., & Zhou, Y. (2025). AI-empowered genome decoding: Applications of large language models in genomics. *Frontiers of Digital Education, 2*(1), 14.
38. Yiu, T., Chen, B., Wang, H., Feng, G., Fu, Q., & Hu, H. (2025). Transformative advances in single-cell omics: A comprehensive review of foundation models, multimodal integration and computational ecosystems. *Journal of Translational Medicine, 23*(1), 1176.
39. Zhou, C., Brown, M., Blaxter, M., Darwin Tree of Life Project Consortium, McCarthy, S. A., & Durbin, R. (2025). Oatk: a de novo assembly tool for complex plant organelle genomes. *Genome Biology, 26*(1), 235.
40. Wang, Z., Wang, Z., Jiang, J., Chen, P., Shi, X., & Li, Y. (2025, July). Large language models in bioinformatics: A survey. In *Findings of the Association for Computational Linguistics: ACL 2025* (pp. 3602–3615).
41. Sokhansanj, B. A., & Rosen, G. L. (2025). Regulating genome language models: Navigating policy challenges at the intersection of AI and genetics. *Human Genetics, 144*(9), 949–970.
42. DaSilva, L. F., Senan, S., Kribelbauer-Swietek, J. F., Patel, Z. M., Louis, L. K., Reddy, A. J., ... & Pinello, L. (2025). Designing synthetic regulatory elements using the generative AI framework DNA-diffusion. *Nature Genetics*, 1–15.
43. Su, X., Li, X., Lin, Y., Xie, Z., Zhi, D., & Ji, S. (2025). *Language models for controllable DNA sequence design.* arXiv preprint arXiv:2507.19523.
44. Wu, W., Li, Q., Zhang, Y., Zhan, Z., Chen, R., Li, M., ... & Wang, Z. (2025). *GENERator: a long-context generative genomic foundation model.* arXiv preprint arXiv:2502.07272.
45. Sadeghi, E., Deng, X., Lin, I., Copp, S. M., & Bogdanov, P. (2025). Property-isometric variational autoencoders for sequence modeling and design. arXiv preprint arXiv:2509.14287.
46. Lee, D. H., Park, E. G., Lee, Y. J., Jeong, H. S., Roh, H. Y., Jeong, G. R., … & Kim, H. S. (2025). Advances in Artificial Intelligence (AI) models and generative algorithms represent a new paradigm for genomics research. *International Journal of Molecular Sciences, 26*(22), 10925.
47. Ekambaram, S., & Dokholyan, N. V. (2026). Peptide-based drug design using generative AI. *Chemical Communications, 62*(3), 672–691.
48. Li, Z., Ni, Y., Beardall, W. A., Xia, G., Das, A., Stan, G. B., & Zhao, Y. (2024). Discdiff: Latent diffusion model for DNA sequence generation. arXiv preprint arXiv:2402.06079.
49. Miller, C., Portlock, T., Nyaga, D. M., & O'Sullivan, J. M. (2024). A review of model evaluation metrics for machine learning in genetics and genomics. *Frontiers in Bioinformatics, 4*, 1457619.
50. Xie, Y., Peng, G., Tikhonova, I., Enns, G., Zhao, H., Cowan, T., & Scharfe, C. (2025). Improving newborn screening accuracy through genome sequencing, targeted metabolomics, and machine learning. *BMC Medical Genomics, 18*(1), 187.
51. Salazar-Garcés, L. F., Morales-Urrutia, E., Cashabamba, F., Proaño Alulema, R. X., & Leiva Suero, L. E. (2025). Evaluating AI-driven precision oncology for breast cancer in low-and middle-income countries: A review of machine learning performance, genomic data use, and clinical feasibility. *Frontiers in Digital Health, 7*, 1702339.

52. Hasanzadeh, F., Josephson, C. B., Waters, G., Adedinsewo, D., Azizi, Z., & White, J. A. (2025). Bias recognition and mitigation strategies in artificial intelligence healthcare applications. *NPJ Digital Medicine, 8*(1), 154.
53. Afreen, J., Mohaghegh, M., & Doborjeh, M. (2025). Systematic literature review on bias mitigation in generative AI. *AI and Ethics, 5*(5), 4789–4841.

Application of AI in Genomics 4

The rapid expansion of high-throughput sequencing technologies has transformed genomics into a data-intensive scientific discipline, generating unprecedented volumes of heterogeneous biological data. While these advances have enabled large-scale studies of genetic variation, gene regulation, microbial ecosystems, and population diversity, they have also introduced substantial analytical challenges. Sequencing noise, high dimensionality, nonlinear biological interactions, and population heterogeneity often exceed the capacity of traditional statistical and rule-based computational approaches. In this context, artificial intelligence (AI) has emerged as a transformative methodological framework for genomic analysis.

AI methods encompassing machine learning, deep learning, probabilistic modeling, and graph-based approaches offer the capacity to learn complex patterns directly from data while integrating diverse sources of biological information. Unlike conventional pipelines that rely on handcrafted features or rigid assumptions, AI-driven models can adapt to varying data modalities, capture nonlinear relationships, and scale to population-level datasets. These capabilities make AI particularly well suited to addressing foundational challenges across genomics, from raw sequence interpretation to functional and evolutionary inference.

This chapter explores the applications of AI across the genomic analysis pipeline, emphasizing both biological intuition and methodological relevance. Beginning with AI-assisted base calling and genome assembly, the chapter examines how learning-based approaches improve sequence accuracy and assembly completeness, particularly in long-read and repeat-rich genomic contexts. It then considers AI-driven variant calling and prioritization, highlighting how probabilistic models enhance sensitivity, integrate functional context, and support discovery in complex disease and non-coding genomics.

K. Shaikh et al., *Artificial Intelligence in Genomics*, Synthesis Lectures on Engineering, Science, and Technology, https://doi.org/10.1007/978-3-032-25486-3_4

Subsequent sections address AI-based modeling of gene expression and epigenomic regulation, where multi-scale architectures integrate sequence, chromatin, and three-dimensional genome organization to elucidate regulatory mechanisms. The chapter further examines AI-driven disease risk prediction and polygenic scoring, focusing on research-oriented applications that model complex trait architecture while acknowledging ethical and population-specific limitations. Expanding beyond the human genome, the chapter discusses AI applications in microbiome and metagenomic analysis, where learning-based methods enable scalable interpretation of diverse microbial communities. Finally, the role of AI in population genomics is explored, illustrating how integrative models contribute evolutionary insights while complementing classical population genetic theory.

Throughout the chapter, emphasis is placed on responsible AI deployment, transparent benchmarking, and biological interpretability. Rather than presenting AI as a replacement for experimental or theoretical frameworks, this chapter positions AI as a complementary tool that enhances hypothesis generation, accelerates discovery, and supports translational genomics. By providing a structured overview of AI applications across genomic domains, this chapter aims to equip researchers, clinicians, and decision-makers with a principled understanding of how AI is reshaping the future of genomic science.

4.1 AI-Assisted Genome Assembly and Base Calling

Genome assembly and base calling represent the earliest and most critical stages of genomic data analysis. Base calling translates raw sequencing signals into nucleotide sequences, while genome assembly reconstructs long genomic regions or entire genomes from fragmented reads. Errors introduced at these stages propagate throughout downstream analyses, influencing variant detection, functional annotation, comparative genomics, and biological interpretation. Consequently, the accuracy and reliability of early-stage genomic processing directly determine the quality of subsequent analytical outcomes.

As sequencing technologies have diversified, particularly with the widespread adoption of long-read platforms, the limitations of rule-based and heuristic-driven algorithms have become increasingly apparent. Long-read sequencing technologies generate complex, noisy signals that encode rich biological information but challenge conventional probabilistic models. Artificial intelligence (AI) has therefore emerged as a central methodological advancement for improving accuracy, scalability, and robustness in genome assembly and base calling [1, 2].

AI-driven approaches differ fundamentally from traditional methods by learning complex relationships directly from data rather than relying on predefined statistical assumptions. This capability is especially valuable in genomics, where signal variability arises from both biological complexity and technical noise. The integration of AI into early-stage genomic processing has reshaped how raw molecular information is transformed into analyzable genomic sequences, enabling more faithful representations of genome structure across diverse organisms.

To understand the role of AI in genome reconstruction, it is essential to first examine its impact on base calling, the point at which biological signals are initially interpreted.

4.1.1 AI-Based Base Calling: Biological Signals to Nucleotide Sequences

- **Intuition and Biological Relevance**

Base calling is the computational inference of nucleotide identities from raw sequencing signals, such as fluorescence intensities in sequencing-by-synthesis platforms or electrical current changes in nanopore-based technologies. Biologically, these signals are influenced not only by the nucleotide being measured but also by neighboring bases, molecular conformation, chemical modifications, and instrument-specific effects. As a result, the relationship between signal and sequence is inherently nonlinear and context dependent.

Traditional base callers rely on probabilistic models, hidden Markov frameworks, or signal-thresholding heuristics. While effective for earlier short-read technologies, these approaches struggle to capture high-dimensional dependencies present in long-read and high-throughput sequencing data. Errors frequently arise in homopolymer regions, repetitive sequences, and low-complexity genomic contexts.

AI-based base callers address this challenge by learning nonlinear mappings between signal patterns and nucleotide sequences directly from large-scale training data. Deep learning architectures—including convolutional neural networks, recurrent neural networks, and transformer-based models—can capture both local signal features and long-range contextual dependencies. This is biologically meaningful because nucleotide identity is influenced by sequence context, especially in regions affected by epigenetic modifications or structural constraints [2].

By modeling signal dependencies across extended windows, AI systems effectively approximate the biochemical processes underlying sequencing, rather than relying on simplified assumptions.

- **Practical Implications in Genomics**

In long-read sequencing, AI-assisted base calling has significantly reduced systematic errors such as insertions and deletions in homopolymer regions. This improvement is particularly important for accurate gene annotation, as small indel errors can disrupt open reading frames and affect downstream functional predictions.

In microbial genomics, enhanced base calling accuracy improves the identification of coding sequences, operon structures, and antimicrobial resistance genes. In cancer genomics, improved signal interpretation reduces false-positive variant calls in low-complexity regions, thereby increasing confidence in somatic mutation analysis and tumor heterogeneity studies.

AI models can also be trained to recognize signal deviations caused by chemically modified bases, enabling simultaneous sequence and epigenetic inference without additional experimental protocols. Although these applications remain primarily research-oriented, they illustrate the broader biological value of AI-driven signal interpretation and its potential to unify sequence and epigenomic analysis pipelines [3].

Once base calling produces nucleotide reads, the next challenge lies in assembling these fragments into coherent genomic structures.

4.1.2 AI-Enhanced Genome Assembly

- **Assembly Challenges and Biological Complexity**

Genome assembly aims to reconstruct the original genome from overlapping reads generated during sequencing. This task is complicated by biological features such as repetitive elements, segmental duplications, gene families, and structural variation. Many of these features play important functional roles, including regulatory control and genome evolution, yet they create ambiguity during computational reconstruction.

Technical factors, including sequencing errors, uneven coverage, and read length variability—further exacerbate assembly challenges. Classical assembly approaches rely on graph-based representations, such as overlap-layout-consensus or de Bruijn graphs, combined with heuristic thresholds. While effective in many contexts, these heuristics may oversimplify biologically complex regions.

AI-enhanced assembly approaches augment traditional methods by learning to resolve ambiguities arising from both biological repetition and sequencing noise. Neural models can be trained to distinguish true overlaps from spurious alignments, a task that is particularly difficult in repeat-rich genomes and polyploid organisms [4].

- **AI in Consensus Refinement and Hybrid Assembly**

Beyond reading overlap resolution, AI models play a critical role in consensus refinement, where assembled contigs are polished to correct systematic sequencing errors. Unlike rule-based correction methods, AI-based polishers learn platform-specific and context-dependent error profiles, allowing them to adapt across sequencing technologies and datasets.

Table 4.1 Roles of AI in genome assembly pipelines

Assembly stage	Biological challenge	AI contribution
Read overlap detection	Repeats, duplications	Overlap classification
Graph simplification	Structural ambiguity	Learning-based pruning
Consensus generation	Sequencing errors	Error profile learning
Hybrid integration	Data heterogeneity	Adaptive reconciliation

Hybrid assembly frameworks, which integrate short- and long-read data, particularly benefit from AI-guided error correction and alignment scoring. Short reads provide high base-level accuracy, while long readings resolve structural complexity; AI models help reconcile these complementary data types into coherent assemblies.

In large eukaryotic genomes, AI-assisted assembly has enabled more complete reconstruction of regulatory regions, centromeric sequences, and structural variants. These improvements support studies in evolutionary genomics, comparative biology, and functional annotation, enhancing biological interpretability rather than merely improving computational performance metrics [5]. Table 4.1 shows roles of AI in genome assembly pipelines and graphical representation of roles of AI in this area shown in Fig. 4.1.

4.1.3 Scalability, Limitations, and Responsible Use

AI-based base calling and assembly models are computationally intensive during training but highly scalable during deployment. Pre-trained models can be fine-tuned for new organisms or sequencing chemistries, which is particularly valuable for non-model species and biodiversity genomics. However, model generalization remains a challenge. Biases in training data may affect underrepresented genomic contexts, organisms, or sequencing conditions. Interpretability and validation are also critical concerns, as AI models output probabilistic predictions that must be carefully assessed.

Importantly, these systems refine inference rather than replace experimental validation and should not be interpreted as producing clinically approved results. Transparent benchmarking, version control, and responsible deployment are essential to ensure reproducibility and trust in AI-assisted genomic pipelines [6]. With assembled genomes as a foundation, AI further contributes to genomics through the identification and interpretation of genetic variation.

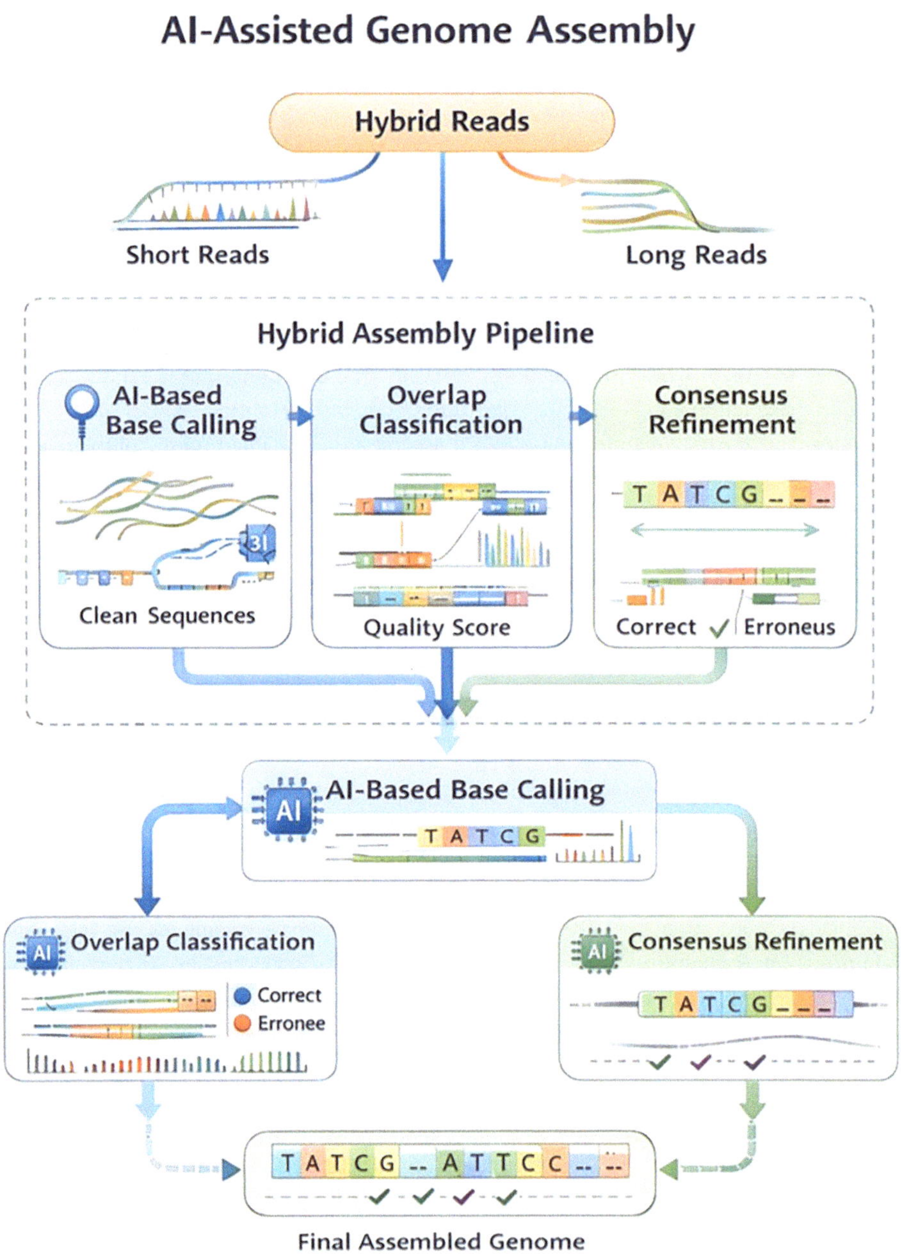

Fig. 4.1 Conceptual overview of an AI-assisted hybrid genome assembly pipeline, illustrating AI-based base calling, overlap classification, and consensus refinement leading to a final assembled genome

4.2 Variant Calling and Prioritization

Variant calling and prioritization are central to extracting biological and translational values from genomic data. Variant calling identifies differences between an individual genome and a reference genome, while variant prioritization assesses the potential functional, regulatory, or phenotypic relevance of these differences. Genomic variation spans single-nucleotide polymorphisms (SNPs), insertions and deletions (indels), copy number variations, and structural rearrangements, all of which contribute to phenotypic diversity, disease susceptibility, and drug response. Accurately identifying and interpreting this variation is therefore fundamental to genomics research and precision medicine.

As sequencing datasets have increased in scale, depth, and complexity, traditional rule-based pipelines often relying on fixed thresholds and handcrafted features have struggled to balance sensitivity and specificity, particularly for rare variants, low-frequency somatic mutations, or variants in complex genomic regions. Repetitive elements, GC-rich loci, and structurally variable regions pose significant challenges for conventional approaches, resulting in false positives or false negatives that propagate downstream.

AI has emerged as a powerful approach to improve both variant detection and interpretation by learning discriminative patterns directly from data rather than relying solely on predefined rules [7, 8]. Machine learning models can integrate multiple layers of information sequencing signals, alignment quality, genomic context, and multi-omic features to enhance the fidelity of variant identification and prioritize variants based on their predicted biological impact. Understanding AI's contribution begins with a detailed examination of how it improves variants, calling itself the first critical step in the workflow.

4.2.1 AI in Variant Calling: Detection and Classification

- **Intuition and Biological Context**

Variant calling requires distinguishing true genetic variants from sequencing errors, alignment artifacts, and platform-specific noise. This is inherently complex, as genuine variants can occur in low-coverage regions, within repetitive sequences, or in structurally diverse loci. For example, indels in homopolymer tracts or complex rearrangements in cancer genomes can be difficult to detect reliably using traditional statistical thresholds. Moreover, biological signals are context-dependent: the same nucleotide change may be more likely in one genomic region than another due to local sequence context or chromatin accessibility.

AI-based variant callers address these challenges by learning patterns from raw and processed sequencing data. Deep learning architectures including convolutional, recurrent, and attention-based networks can integrate base quality scores, read pileups, alignment context, and signal-level features to differentiate true variants from noise. By learning these complex dependencies, AI models can capture subtle cues that rule-based algorithms may overlook, improving sensitivity for small indels, multi-nucleotide polymorphisms, and variants in difficult genomic regions [8].

From a biological perspective, this capability allows variant calling to more accurately reflect the underlying genome, rather than producing results constrained by arbitrary thresholds. AI models can also output probabilistic confidence scores for each variant, which reflect the uncertainty inherent in low-coverage or ambiguous regions. These confidence scores are particularly valuable when integrating variant calls into downstream analyses such as population genetics or rare disease studies.

- **Practical Genomic Applications**

  AI-enhanced variant calling has wide-ranging applications.
  - **Population Genomics**: Consistent and accurate detection of variants across large cohorts reduces batch effects and technical bias, supporting studies of allele frequency, population structure, and evolutionary inference.
  - **Cancer Genomics**: AI models improve detection of low-frequency somatic mutations in heterogeneous tumor samples, aiding in the characterization of tumor evolution and subclonal architecture.
  - **Microbial Genomics**: For pathogen sequencing, AI-based callers help identify rare resistance mutations that may be overlooked by standard pipelines.
  - **Structural Variant Detection**: Learning-based models can integrate signals from read-depth, split readings, and long-read alignment patterns to detect deletions, duplications, and inversions with improved specificity.

Table 4.2 shows various types of inputs for AI -based variant calling.

AI-enhanced variant calling has wide-ranging applications across genomic datasets (Table 4.3). The table summarizes common variant types, the biological and computational challenges they present, limitations of traditional detection, and the contributions of AI-based methods. Identifying variants accurately is only the first step; understanding which variants matter biologically requires robust prioritization strategies.

Table 4.2 Inputs for AI-based variant calling

Input type	Description	Biological relevance
Base quality scores	Per-base sequencing confidence	Weight contribution of each read
Read alignment features	Mapping quality, soft clipping	Detect artifacts vs true variants
Local sequence context	Neighboring nucleotides	Context-dependent error patterns
Multi-omic annotations (optional)	Epigenetic marks, conservation	Prioritize biologically relevant variants

Table 4.3 Variant types and AI detection complexity

Variant type	Biological/genomic challenge	Traditional detection limitations	AI contribution
Single nucleotide polymorphisms (SNPs)	Often occur in repetitive or GC-rich regions; subtle signal differences	Hard to distinguish from sequencing errors in low-coverage regions	Learns subtle base-quality patterns and sequence context for high-confidence detection
Insertions/ deletions (indels)	Small indels in homopolymer regions; longer indels disrupt alignment	Misalignment leads to false positives/negatives	Context-aware models reduce indel miscalling and integrate read-pair information
Copy number variations (CNVs)	Dosage differences in repeated or segmentally duplicated regions	Signal thresholds can misclassify true CNVs	AI integrates coverage depth, alignment signatures, and local context for robust detection
Structural variants (SVs)	Inversions, translocations, large deletions/duplications; complex breakpoint junctions	Limited by short-read data or heuristic filters	Learning-based approaches combine read-depth, split-reads, and long-read evidence to detect complex rearrangements
Rare/ low-frequency variants	Variants present in small subpopulations or tumor subclones	Underpowered in traditional pipelines; easily missed	AI leverages probabilistic models and cohort-level patterns to improve sensitivity
Non-coding/ regulatory variants	Often functionally subtle; lack of coding impact	Hard to prioritize; scoring functions incomplete	Integrates epigenomic, conservation, and functional annotations for probabilistic prioritization

4.2.2 AI-Driven Variant Prioritization

Genomes contain millions of variants, the majority of which are functionally neutral. Prioritization aims to identify variants most likely to impact gene function, regulation, or phenotype. Traditional scoring systems rely on evolutionary conservation, predicted protein effects, or population frequency thresholds. However, these approaches may fail to integrate diverse sources of biological evidence or capture subtle combinatorial patterns.

AI models excel at integrating multi-dimensional data for variant prioritization. By learning complex relationships among features such as sequence context, functional annotations, regulatory impact, evolutionary conservation, and tissue-specific expression, AI can produce probabilistic rankings that highlight variants with high potential biological significance [9]. This is particularly useful in rare disease studies or precision medicine applications, where the causal variant may be unique to a single individual.

Figure 4.2 shows conceptual overview of AI-driven variant calling and prioritization, integrating sequencing data, functional annotations, evolutionary conservation, and multi-omic features. While AI improves both variant detection and prioritization, its effectiveness depends on data quality, interpretability, and ethical deployment.

- **Few Practical Examples are as follows:**
  - **Rare Disease Genomics**: AI-driven prioritization can combine patient-specific phenotypic information with genomic data to narrow down candidate variants, reducing the burden of experimental validation.
  - **Pharmacogenomics**: AI models help identify variants likely to affect drug metabolism, transport, or target binding, supporting personalized treatment strategies.
  - **Cancer Research**: Prioritization frameworks can highlight somatic mutations that are more likely to disrupt oncogenes or tumor suppressor genes.
  - **Functional Genomics**: By integrating multi-omic datasets, AI can rank non-coding variants for regulatory potential, facilitating hypothesis generation for experimental studies.

4.2.3 Challenges, Bias, and Responsible Interpretation

Despite their power, AI-based variant calling and prioritization models are sensitive to biases in training data. Population representation is a major concern: variants common in underrepresented populations may be systematically misclassified, limiting model generalizability. Annotation completeness, sequencing platform differences, and cohort-specific artifacts also affect model performance.

Interpretability is another critical factor. Researchers must understand why a variant is prioritized to trust its biological relevance. AI outputs should be treated as probabilistic decision-support tools rather than definitive conclusions. Transparency, benchmarking, and external validation are essential to ensure responsible use in research contexts [10].

Ethical considerations are also paramount. In translational settings, misinterpretation of AI-prioritized variants could misguide research or therapeutic decision-making. Therefore, AI should complement—not replace—experimental validation and expert interpretation.

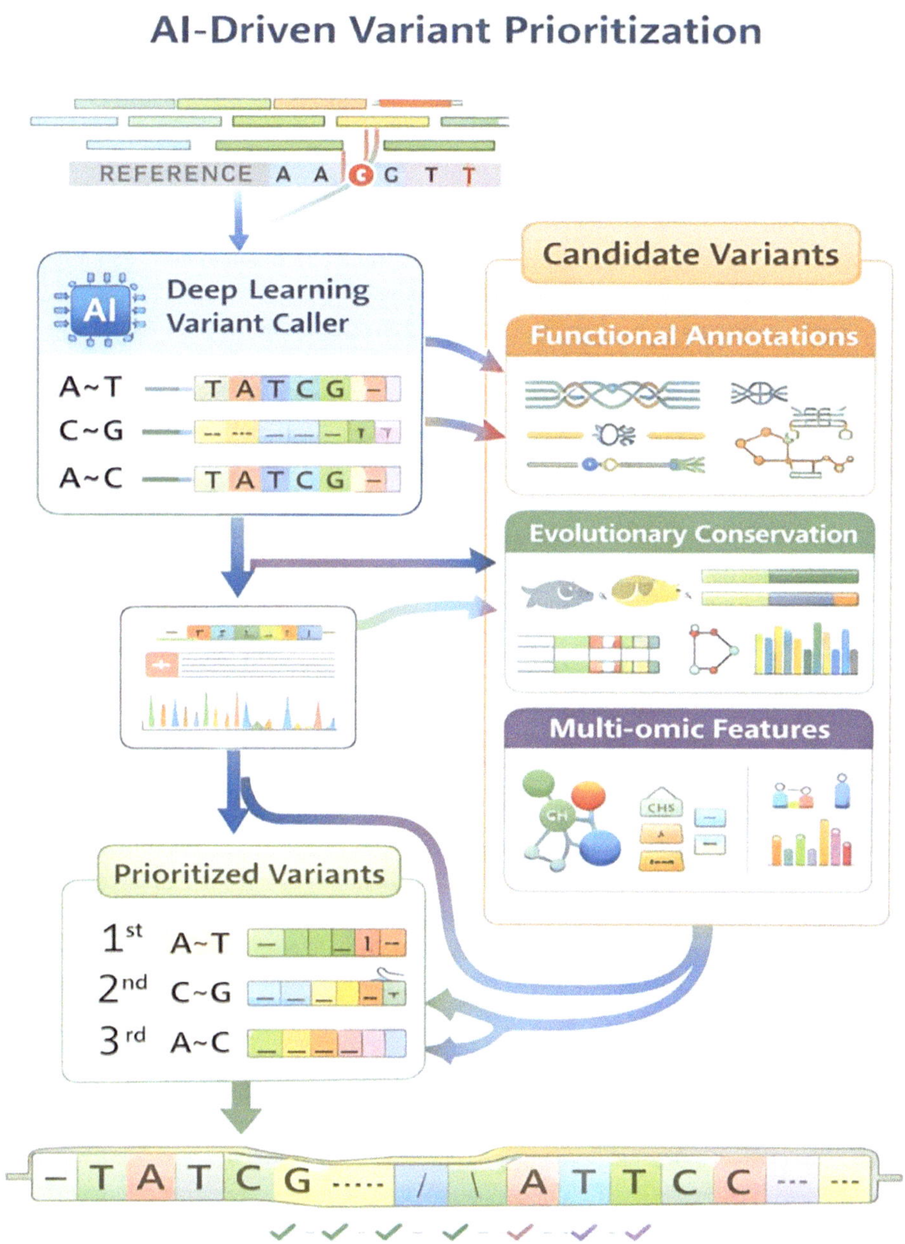

Fig. 4.2 Conceptual overview of AI-driven variant calling and prioritization integrating sequencing data, functional annotations, evolutionary conservation, and multi-omic features

4.3 Predicting Gene Expression and Regulatory Patterns

Gene expression represents the primary mechanism through which genomic information is translated into cellular phenotype. While the genome provides a largely static blueprint, gene expression is highly dynamic and context-dependent, influenced by regulatory elements, epigenomic states, chromatin structure, and environmental cues. Accurately predicting gene expression from genomic and regulatory information has long been a central challenge in computational biology, as it requires modeling the nonlinear, combinatorial, and context-specific interactions that control transcription.

Artificial intelligence (AI) has emerged as a powerful framework for addressing these challenges. Unlike traditional linear or probabilistic models, AI can capture hierarchical and nonlinear dependencies, integrating information from DNA sequence motifs, chromatin accessibility, transcription factor occupancy, and three-dimensional genome organization [11, 12]. AI-driven prediction does not replace experimental assays such as RNA-seq or single-cell transcriptomics; instead, it complements them by providing hypothesis-generating insights, enabling prioritization of regulatory variants, and extrapolating expression patterns across untested tissues or conditions. To appreciate AI's impact, it is first necessary to understand the biological complexity underpinning gene expression regulation.

4.3.1 Biological Foundations of Gene Expression Regulation

Gene expression is orchestrated through a multi-layered regulatory system. Promoters, enhancers, silencers, transcription factors, and co-regulators collectively determine the transcriptional output of each gene. Chromatin accessibility and histone modifications provide additional regulatory layers, while three-dimensional chromatin organization enables distal regulatory elements to interact with promoters via looping mechanisms.

From a biological perspective, expression levels result from the integration of multiple weak regulatory signals rather than a single dominant factor. Traditional statistical models often fail to capture this combinatorial logic, particularly when distal or context-dependent interactions are involved. AI models, particularly deep learning architectures, are well suited to this challenge because they can learn hierarchical representations: low-level features such as transcription factor binding motifs, mid-level features representing regulatory modules, and high-level features corresponding to gene-level expression patterns [12].

This biological complexity motivates the use of AI architectures capable of learning multi-scale representations, capturing both local sequence dependencies and long-range regulatory interactions.

Table 4.4 Common AI inputs for gene expression prediction

Input type	Biological meaning	Example use
DNA sequence	Encodes regulatory motifs	Predict promoter strength, transcription factor binding
Chromatin accessibility	Indicates regulatory activity	Predict cell-type–specific expression
Transcription factor binding	Reflects regulatory control signals	Integrate motif occupancy for expression modeling
3D genome contacts	Enables long-range regulation	Model enhancer–gene interactions

4.3.2 AI Architectures for Gene Expression Prediction

AI models for predicting gene expression integrate genomic sequence features, regulatory annotations, and contextual information. Convolutional neural networks (CNNs) are widely used to detect local sequence motifs, such as transcription factor binding sites, nucleosome positioning signals, or CpG islands. Attention-based and transformer architectures enable modeling of long-range interactions, reflecting the biological reality that distal enhancers and chromatin loops play critical roles in transcriptional control. Graph Neural Networks (GNNs) are increasingly applied to integrate regulatory networks and three-dimensional genome organization [13]. In GNN-based models, nodes represent genes, regulatory elements, or chromatin regions, while edges encode functional, spatial, or co-expression relationships. These networks can capture combinatorial regulatory logic and provide biologically interpretable predictions of expression changes across cell types or conditions. Table 4.4 shows the common AI inputs for gene expression prediction.

Beyond architecture and inputs, model performance is evaluated by the ability to generalize across tissues, cell types, and experimental conditions.

4.3.3 Practical Applications in Genomics

AI-based gene expression prediction provides practical value across multiple genomic research domains:

- **Functional Genomics**: These models can prioritize regulatory variants in non-coding regions by estimating their impact on transcriptional output, enabling researchers to focus on potentially functional mutations.
- **Developmental Biology**: AI helps model temporal gene expression trajectories during differentiation, identifying regulatory programs governing cell fate transitions.

- **Disease Genomics**: In cancer, AI models predict dysregulated gene expression associated with oncogenic rewiring, highlighting candidate drivers and potential therapeutic targets.
- **Population Genomics**: AI facilitates interpretation of expression quantitative trait loci (eQTLs) without requiring expression data for every tissue or condition, enabling scalable analysis across large cohorts [14].
- **Pharmacogenomics**: Predicted expression changes can help anticipate drug response or adverse effects by linking regulatory variants to gene expression patterns in relevant tissues.

Figure 4.3 highlights multi-scale learning from motif-level to gene-level representations, capturing both local and distal regulatory influences.

4.3.4 Challenges, Bias, and Responsible Interpretation

Despite their predictive power, AI-based gene expression models are subject to limitations:

- **Dataset Bias**: Overrepresentation of well-studied tissues, cell types, or regulatory features can skew predictions. Models may perform poorly in underrepresented contexts.
- **Interpretability**: Deep learning models are often "black boxes," making it challenging to link specific input features to predicted expression patterns. Approaches such as feature attribution and attention mapping can partially address this issue.
- **Causality versus Correlation**: AI predictions indicate probable regulatory influence but do not establish mechanistic causation. Experimental validation remains essential.
- **Ethical Considerations**: Predictive models should be applied in research contexts and not used for clinical or diagnostic decision-making without rigorous validation [15].

Addressing these challenges requires transparent benchmarking, careful cross-tissue validation, and integration with experimental data.

4.3.5 Epigenomic Modeling (Methylation and Chromatin States)

Epigenomic regulation governs how the same genomic sequence can give rise to diverse cellular phenotypes. DNA methylation, histone modifications, and chromatin accessibility collectively shape gene expression without altering nucleotide sequences. These layers of regulation are critical for controlling developmental processes, maintaining cellular identity, responding to environmental stimuli, and mediating disease states. However, the

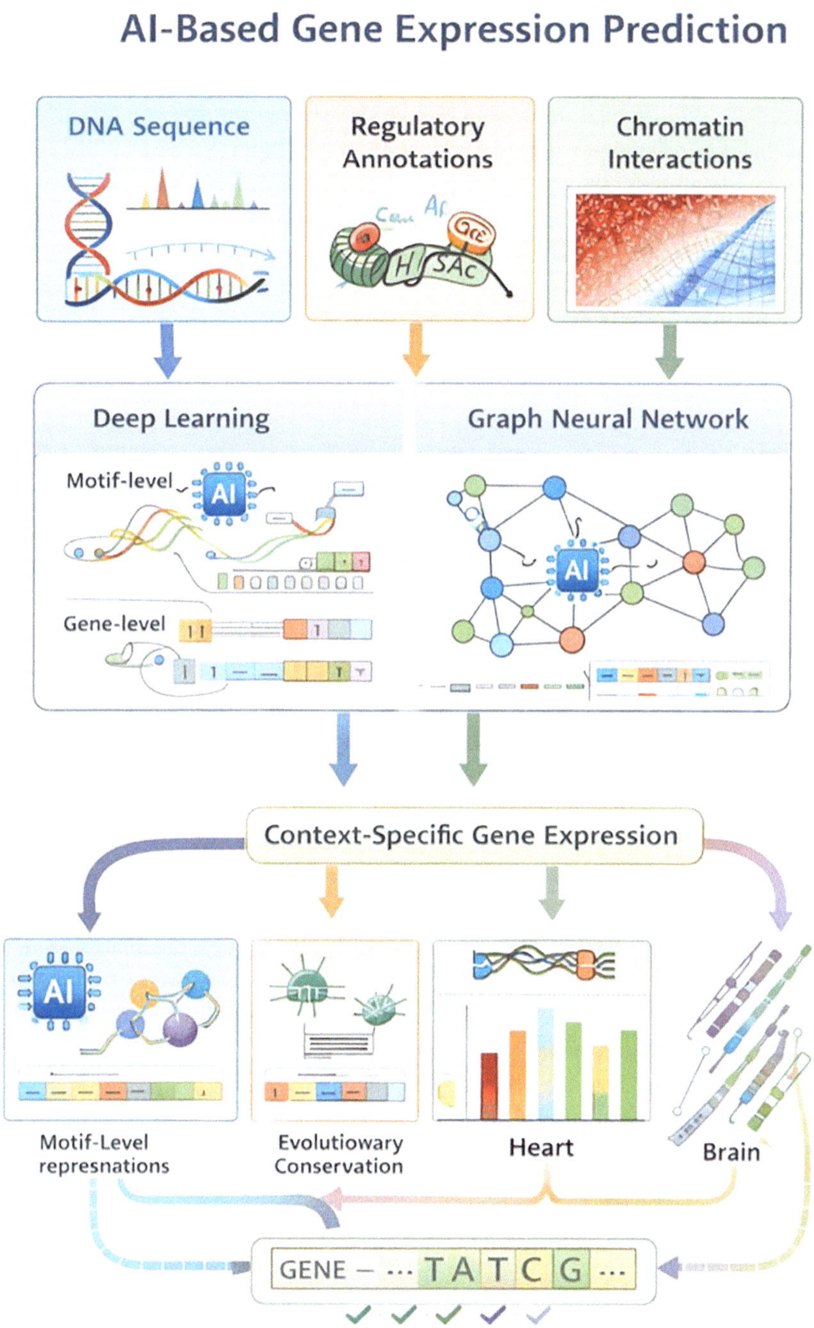

Fig. 4.3 AI-based gene expression prediction using multi-scale regulatory modeling

high dimensional, heterogeneous, and context-dependent nature of epigenomic signals presents major challenges for conventional statistical or rule-based analysis pipelines.

Artificial intelligence (AI) has become central to epigenomic analysis due to its ability to integrate multi-modal, high-dimensional datasets and to uncover latent regulatory patterns that are difficult to detect using conventional approaches [13, 16]. By learning complex, nonlinear relationships across multiple epigenomic layers, AI can generate predictive models for methylation states, chromatin accessibility, histone modifications, and overall chromatin state annotations. Such models complement experimental assays, enabling scalable predictions across cell types, tissues, and conditions where experimental data may be sparse. Understanding AI's role in epigenomics requires an appreciation of the biological complexity of epigenetic regulation.

4.3.6 Biological Basis of Epigenomic Regulation

Epigenetic regulation involves dynamic modifications to DNA and chromatin that influence gene expression without changing the underlying nucleotide sequence. DNA methylation, typically occurring at cytosine residues in CpG dinucleotides, can repress transcription by obstructing transcription factor binding or recruiting repressive complexes. Histone modifications, including acetylation, methylation, phosphorylation, and ubiquitination, define chromatin states such as active, poised, or repressed regulatory regions. Chromatin accessibility determines whether regulatory proteins can physically access DNA and is highly cell-type specific.

These mechanisms are inherently dynamic, context-dependent, and reversible. For example, during differentiation, stem cells progressively establish lineage-specific methylation patterns and chromatin states, while cancer cells often exhibit aberrant methylation and chromatin remodeling that drives oncogenic gene expression programs. Epigenomic signals are also noisy due to experimental variability, sequencing biases, and technical limitations of assays such as bisulfite sequencing or ATAC-seq. Traditional linear or threshold-based models struggle to capture the combinatorial and context-specific relationships among these epigenomic marks [16].

AI models, particularly deep learning architectures, are well suited to capture these distributed, high-dimensional regulatory patterns. By learning latent representations that reflect combinations of regulatory features, AI approaches can infer epigenetic states even in partially observed datasets, providing insights into regulatory landscapes that would be difficult to obtain experimentally. This biological motivation has driven the development of specialized AI models for integrating diverse epigenomic datasets.

Table 4.5 Epigenomic features modeled by AI

Epigenomic layer	Biological role	AI modeling objective
DNA methylation	Gene repression/activation	Predict methylation profiles in specific cell types
Histone marks	Define chromatin states (active, repressed, poised)	Classify chromatin states and regulatory regions
Chromatin accessibility	Regulatory potential of DNA regions	Detect accessible promoters/ enhancers
Chromatin states	Integrated regulatory context	Generate functional annotations for genomic regions

4.3.7 AI Approaches to Epigenomic Modeling

AI methodologies for epigenomic modeling are diverse, ranging from single-task deep learning networks to multi-modal integrative frameworks:

- **Deep Learning for Single-Layer Predictions**: Convolutional neural networks (CNNs) are used to predict local patterns such as methylation status or nucleosome positioning from DNA sequence. Recurrent neural networks (RNNs) and attention-based transformers model broader genomic context, capturing dependencies across distal regulatory elements.
- **Multi-Modal Integration**: Modern AI frameworks combine methylation, histone marks, and chromatin accessibility into unified embeddings, enabling predictions of regulatory states in conditions or cell types where direct experimental measurements are lacking [17]. These multi-modal models support comparative epigenomics and facilitate the study of development, disease, and evolutionary epigenetics.
- **Graph-Based and Network Models**: Epigenetic interactions are often nonlinear and involve multiple regulatory nodes. Graph Neural Networks can represent chromatin regions, genes, and regulatory elements as nodes connected by functional or spatial edges, allowing AI to model complex regulatory architectures.

Table 4.5 shows epigenomic feature modeled by AI in the genomic.

The predictive power of AI models enables practical applications across diverse areas of genomic research.

4.3.8 Practical Applications in Genomics

AI-based epigenomic modeling has numerous applications in both research and translational genomics:

- **Cancer Genomics**: AI predicts aberrant methylation patterns and histone modification changes in tumors, enabling the identification of silenced tumor suppressor genes or activated oncogenes. Multi-modal models can link these epigenomic alterations to transcriptional dysregulation, supporting biomarker discovery and mechanistic studies [15].
- **Developmental Biology**: During differentiation, AI models can reconstruct temporal trajectories of methylation and chromatin states, identifying stage-specific regulatory programs. This informs understanding of lineage commitment and cellular plasticity.
- **Functional Genomics**: By integrating methylation, accessibility, and histone modification data, AI models highlight regulatory regions most likely to influence gene expression, aiding variant interpretation in non-coding regions.
- **Comparative Epigenomics**: AI models trained across multiple tissues or species enable the identification of conserved regulatory elements and dynamic epigenetic patterns, supporting evolutionary and cross-species studies.

Despite these advances, epigenomic AI models face limitations in generalizability, interpretability, and ethical application.

4.3.9 Challenges, Bias, and Ethical Considerations

AI-driven epigenomic models encounter several practical challenges.

- **Data Heterogeneity**: Epigenomic datasets differ in assay type, resolution, coverage, and noise levels. This variability can introduce bias into AI models, particularly when integrating datasets from multiple platforms or laboratories.
- **Context-Specificity**: Epigenetic marks are highly tissue- and cell-type-specific. Models trained on one context may not generalize to others, necessitating careful validation and potential domain adaptation.
- **Interpretability**: Deep learning models often function as "black boxes." Biologically interpretable frameworks, such as attention mechanisms or feature attribution, are essential to relate predictions to underlying regulatory mechanisms.
- **Ethical and Responsible Use**: Epigenomic predictions should be treated as decision-support tools. They are not clinically validated diagnostics, and results must be contextualized with experimental evidence. Consideration of population diversity, data privacy, and responsible interpretation is critical, particularly when models are used in translational research [10].

Integrating epigenomic modeling with gene expression and variant interpretation supports holistic regulatory genomics and multi-layered functional analysis.

4.4 Disease Risk Prediction and Polygenic Scores

Genomic variation underlies individual differences in susceptibility to complex diseases. Unlike monogenic disorders, complex traits such as cardiovascular disease, diabetes, neurodegeneration, and autoimmune disorders result from the cumulative effect of many genetic variants, environmental factors, and their interactions. Polygenic risk scores (PRS) provide a quantitative framework for aggregating the contributions of multiple variants to estimate disease risk in individuals.

Artificial intelligence (AI) has emerged as a transformative tool in disease risk prediction, offering advantages over traditional linear PRS models by incorporating nonlinear interactions, multi-omic data, and environmental context. AI models can integrate genomic, epigenomic, transcriptomic, and clinical information to generate predictive frameworks that are both scalable and biologically informative [18, 19].

To appreciate AI's impact in this domain, it is necessary to understand the biological rationale for polygenic risk modeling and the challenges of conventional approaches.

4.4.1 Biological Rationale and Polygenic Risk

Complex diseases arise from numerous genetic variants, each contributing to a small effect. Single-variant analyses often fail to capture the cumulative risk, particularly for low-effect alleles or interactions among loci. Polygenic risk scores address this limitation by aggregating variant effects, weighted by effect sizes derived from Genome-Wide Association Studies (GWAS) or functional genomics datasets.

From a biological perspective, disease risk reflects both direct gene-disease associations and regulatory networks that influence disease-relevant pathways. Variants in non-coding regions can modulate gene expression or splicing, while epigenetic modifications further mediate gene-environment interactions. Traditional linear models may oversimplify these relationships, assuming additive effects and ignoring higher-order interactions.

AI-driven risk prediction allows modeling of nonlinear and combinatorial effects, incorporating regulatory annotations, gene expression data, and epigenomic information into a unified predictive framework. This capability is particularly valuable in polygenic diseases where risk emerges from the interaction of many small-effect variants [20].

Understanding these biological foundations motivates the design of AI architectures tailored for polygenic risk prediction.

4.4.2 AI Architectures for Disease Risk Prediction

AI models for disease risk prediction range from ensemble machine learning techniques to deep learning networks capable of integrating multi-modal data:

- **Gradient Boosting and Random Forest Models**: These models can capture nonlinear interactions among variants while handling large-scale GWAS datasets. They provide feature importance scores that allow identification of variants with disproportionate influence on predicted risk.
- **Deep Neural Networks**: Fully connected networks and autoencoders can learn latent representations of high-dimensional genotype data, capturing both additive and epistatic effects.
- **Graph Neural Networks (GNNs)**: By representing genes and regulatory elements as nodes within biological networks, GNNs model gene–gene interactions and pathway-level effects on disease risk. Such architectures facilitate biologically interpretable predictions by linking genetic variation to functional networks [21].
- **Multi-modal Integration Frameworks**: Modern AI pipelines combine genotype data with epigenomic profiles, gene expression, metabolomics, and clinical phenotypes. This approach improves risk prediction in scenarios where genetics alone explain a limited fraction of disease variance.

Table 4.6 shows AI models for disease risk prediction.

While AI architectures provide enhanced predictive power, their performance depends critically on data integration, variant weighting, and interpretability.

Table 4.6 AI models for disease risk prediction

AI architecture	Data input	Biological rationale	Practical application
Gradient boosting/ random forest	Genotype, GWAS summary	Nonlinear interactions among variants	Polygenic risk score computation
Deep neural networks	High-dimensional genotype, multi-omic data	Capture latent additive and epistatic effects	Disease susceptibility prediction
Graph neural networks	Gene regulatory networks, variant annotations	Model pathway and network-level interactions	Network-based risk interpretation
Multi-modal AI	Genomics, transcriptomics, epigenomics, clinical traits	Integrate multiple layers influencing risk	Personalized disease risk assessment

4.4.3 Practical Applications in Genomics

AI-driven disease risk prediction has been applied across several domains.

- **Cardiovascular Disease**: Multi-omic AI models have been used to predict individual susceptibility by integrating genomic variants, methylation states, and expression data of genes involved in lipid metabolism, inflammation, and vascular remodeling.
- **Cancer Susceptibility**: AI can combine germline genetic variation with epigenetic and expression profiles to estimate risk for breast, prostate, and colorectal cancers. Network-informed models identify clusters of interacting genes that collectively influence tumor predisposition.
- **Neurodegenerative Disorders**: For conditions such as Alzheimer's and Parkinson's disease, AI models leverage polygenic risk scores alongside epigenomic and transcriptomic data to account for both genetic predisposition and regulatory context.
- **Pharmacogenomics**: Beyond disease risk, AI-enhanced PRS models can help predict adverse drug reactions and therapeutic efficacy, linking individual genetic architecture to treatment response.

Figure 4.4 shows AI-based disease risk prediction: input layers include genomic variants, epigenomic marks, gene expression, and clinical covariates. AI models (e.g., deep neural networks or graph networks) learn latent representations and predict individual-level risk scores, which can then be stratified into low, medium, and high-risk categories. The figure emphasizes multi-modal integration and biological interpretability of predictions.

Despite these applications, AI-based risk prediction faces challenges related to interpretability, bias, and population generalizability.

4.4.4 Challenges, Bias, and Responsible Interpretation

AI models for disease risk prediction are subject to several critical challenges.

- **Population Bias**: Many GWAS and epigenomic datasets overrepresent European ancestry, limiting the generalizability of AI-derived PRS to diverse populations. AI models must be trained on diverse cohorts or adapted through transfer learning to mitigate bias.
- **Data Heterogeneity**: Variations in sequencing quality, variant calling pipelines, and environmental exposure data can influence AI predictions. Harmonization of input data is essential for robust modeling.
- **Interpretability**: Deep learning models often function as black boxes. Biological interpretability can be enhanced by integrating attention mechanisms, feature attribution,

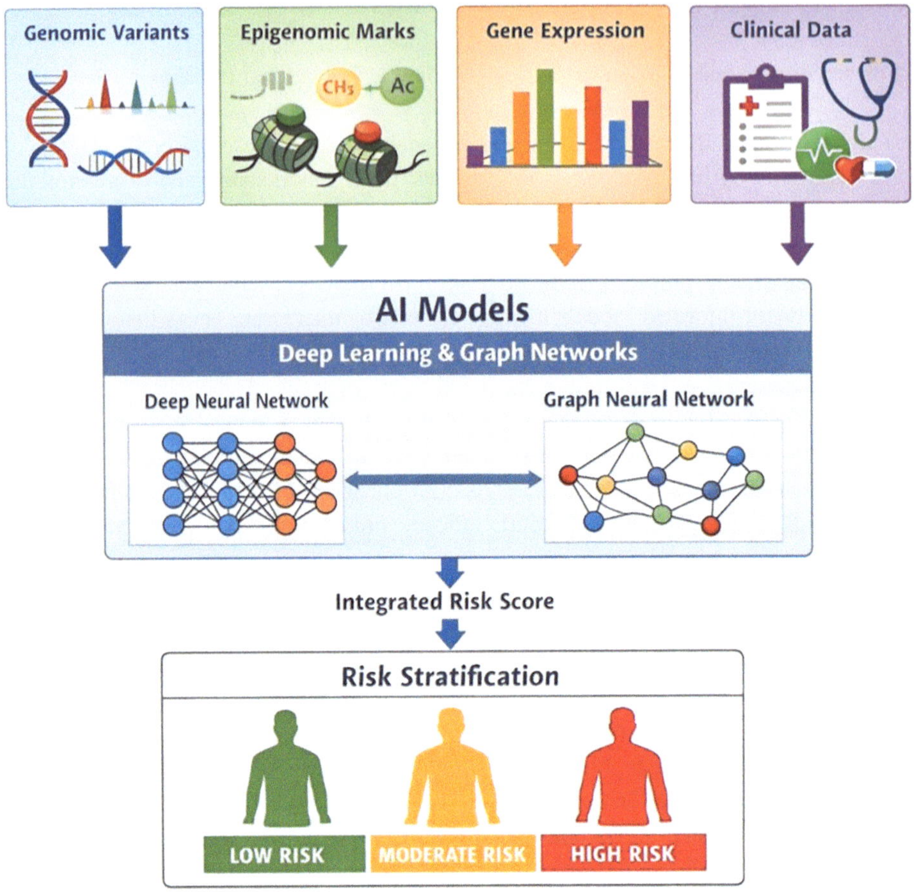

Fig. 4.4 AI-driven disease prediction flowchart

and pathway-level visualization. Understanding which variants and regulatory features drive predictions is crucial for translational relevance [22].

- **Ethical Considerations**: PRS and AI-based predictions carry ethical implications. Risk scores must not be overinterpreted for clinical decision-making, insurance eligibility, or population stratification without rigorous validation. These tools should be used for research and hypothesis generation in precision medicine contexts, not as authoritative clinical diagnostics.

Integrating polygenic risk prediction with functional genomics and epigenomic modeling represents the next frontier for holistic precision medicine.

4.5 Microbiome and Metagenomic Analysis

Table 4.7 shows the AI applications in microbiome and metagenomic analysis.

4.5.1 Biological Background and Motivation

The human microbiome comprises trillions of microorganisms including bacteria, archaea, viruses, and fungi that inhabit diverse ecological niches such as the gut, oral cavity, skin, and respiratory tract. These microbial communities play essential roles in digestion, immune modulation, metabolism, and protection against pathogens. Disruptions in microbial composition and function, commonly referred to as dysbiosis, have been associated with a wide range of complex conditions, including inflammatory bowel disease, metabolic disorders, neurodevelopmental conditions, and cancer susceptibility [23].

Metagenomics enables the culture-independent characterization of microbial communities by directly sequencing DNA extracted from environmental or clinical samples. Unlike traditional genomics, which focuses on a single organism, metagenomic datasets are inherently complex: they consist of mixed genomes, varying abundances, fragmented reads, and incomplete reference coverage. These properties make microbiome analysis a natural application domain for artificial intelligence (AI), which excels at learning patterns from high dimensional, noisy, and heterogeneous data.

AI-driven microbiome analysis aims not only to classify microbial taxa but also to infer functional potential, ecological interactions, and host–microbe associations. As sequencing technologies have matured, the challenge has shifted from data generation to meaningful interpretation, an area where machine learning and deep learning now play a central role.

4.5.2 AI in Taxonomic Profiling and Community Classification

One of the foundational tasks in microbiome research is taxonomic profiling: identifying which microbial taxa are present in a sample and estimating their relative abundances.

Table 4.7 AI applications in microbiome and metagenomic analysis

Task	AI approach	Biological insight
Taxonomic profiling	CNNs, k-mer ML models	Community composition
Functional annotation	Deep protein embeddings	Metabolic potential
Host–microbe integration	Multi-modal deep learning	Interaction mechanisms
Temporal modeling	RNNs, transformers	Microbiome dynamics

Traditional approaches rely on sequence alignment against reference databases, which can be computationally expensive and biased toward well-characterized organisms. AI-based methods instead learn discriminative sequence features directly from raw reads or k-mer representations, enabling faster and more flexible classification [24].

Supervised machine learning models such as random forests and gradient-boosted trees have been widely adopted for microbiome-based phenotype classification, including disease vs. healthy states. More recently, deep learning architectures particularly convolutional neural networks (CNNs) and transformer-inspired sequence models have demonstrated improved robustness to sequencing errors and variable read lengths. These models treat DNA sequences analogously to language, learning embeddings that capture taxon-specific motifs and evolutionary signals.

Practical example: In gut microbiome studies, AI classifiers have been trained to distinguish microbial signatures associated with antibiotic exposure or dietary interventions. Rather than relying on individual taxa, these models learn multivariate patterns across hundreds of microbial features, improving predictive performance while capturing community-level effects.

4.5.3 Functional Metagenomics and Pathway Inference

Beyond identifying which organisms are present, functional metagenomics seeks to determine what metabolic and regulatory capabilities a microbial community possesses. This includes predicting gene families, enzymatic functions, and metabolic pathways encoded within metagenomic assemblies.

AI models contribute at multiple stages of this process. Neural networks trained on annotated protein databases can infer gene function from short, fragmented sequences, even when homology to known genes is weak. Embedding-based models capture biochemical and evolutionary properties of proteins, enabling functional annotation beyond strict sequence similarity [25].

Graph-based learning approaches are increasingly used to model metabolic networks reconstructed from metagenomic data. In this framework, enzymes and metabolites are represented as nodes, while biochemical reactions form edges. AI models can infer missing links, predict pathway activity, and estimate community-level metabolic outputs, such as short-chain fatty acid production in the gut.

Biological relevance: Functional predictions are particularly important in microbiome research because different microbial communities may share similar metabolic functions despite having distinct taxonomic compositions, a phenomenon known as functional redundancy.

4.5.4 Host–Microbiome Interaction Modeling

The biological impact of the microbiome cannot be fully understood without considering its interaction with the host genome, immune system, and environmental exposures. AI enables integrative modeling of host–microbe interactions by combining metagenomic data with host genomics, transcriptomics, metabolomics, and clinical metadata.

Multi-modal deep learning architecture is commonly used for this purpose. Separate subnetworks encode host genetic variants, microbial features, and environmental covariates, which are then fused into a joint representation. Attention mechanisms allow models to highlight specific microbial taxa or pathways that contribute most strongly to host phenotypes [23].

Practical example: In inflammatory disorders, AI models have been used to associate microbial-derived metabolites with host immune gene expression profiles, revealing mechanistic links between microbial activity and inflammatory signaling pathways without asserting causality or clinical decision-making authority.

4.5.5 Temporal and Longitudinal Microbiome Modeling

Microbial communities are dynamic systems that respond to diet, medication, infection, and lifestyle changes over time. Longitudinal microbiome studies generate time-series data that pose additional analytical challenges, including irregular sampling intervals and inter-individual variability.

Recurrent neural networks (RNNs), temporal convolutional networks, and transformer-based time-series models have been adapted to capture microbiome dynamics. These models can learn temporal dependencies, identify stable versus transient taxa, and predict community shifts under perturbations such as antibiotic treatment [26].

From an ecological perspective, AI-driven temporal modeling helps distinguish between resilient core microbiota and opportunistic species. This distinction is essential for understanding microbiome stability and recovery, particularly in early-life development or chronic disease contexts.

4.5.6 Explainability, Bias, and Ethical Considerations

Despite their predictive power, AI models applied to microbiome data face challenges related to interpretability, reproducibility, and bias. Microbiome datasets are highly sensitive to technical factors such as sample collection, DNA extraction protocols, and sequencing platforms. AI models trained on one cohort may not generalize to others without careful normalization and validation.

Explainable AI (XAI) techniques, including feature attribution methods and attention visualization, are increasingly incorporated to identify influential taxa or functional pathways. Such transparency is essential for biological interpretation and for avoiding overstatement of findings, particularly in translational and policy-relevant contexts [27].

Ethical considerations also arise from the use of microbiome data, which can indirectly reveal information about diet, lifestyle, and disease risk. Responsible AI practices emphasize data privacy, population diversity, and clear communication of limitations.

4.6 Population Genomics and Evolutionary Insights

Table 4.8 shows the AI applications in population genomics.

4.6.1 Introduction: Population-Scale Genomics in the Era of AI

Population genomics seeks to understand how genetic variation is distributed within and between populations and how evolutionary forces such as mutation, selection, genetic drift, migration, and recombination shape this variation over time. Advances in high-throughput sequencing have enabled the generation of population-scale genomic datasets encompassing thousands to millions of individuals across diverse ancestries. While these datasets offer unprecedented opportunities to study human history, adaptation, and disease susceptibility, they also pose significant analytical challenges due to their scale, complexity, and heterogeneity.

Artificial intelligence (AI) has emerged as a transformative framework for population genomics by enabling scalable analysis of large datasets, capturing subtle and nonlinear patterns of variation, and integrating diverse sources of genetic and environmental information [28, 29]. AI-driven approaches complement classical population genetic models by emphasizing pattern discovery and predictive inference rather than explicit parametric assumptions. This shift allows researchers to uncover evolutionary signals that may be difficult to detect using traditional methods alone.

Table 4.8 AI applications in population genomics

Task	AI approach	Evolutionary insight
Population structure	Autoencoders, clustering	Ancestry and admixture
Selection detection	Supervised deep learning	Adaptive evolution
Local ancestry	CNN-based models	Chromosomal inheritance
Disease context	Multi-population ML	Health disparities

To appreciate the role of AI in population genomics, it is essential to understand the biological and evolutionary foundations that shape population-level genetic variation.

4.6.2 Biological Foundations of Population Genomics

Genetic variation within populations arises from the accumulation of mutations over generations, filtered by evolutionary forces and demographic history. Population structure, admixture, bottlenecks, and expansions leave characteristic signatures across the genome. These signatures are further influenced by local adaptation, where environmental pressures favor specific genetic variants, and by purifying selection, which removes deleterious alleles.

From a biological perspective, population genomic data are inherently high dimensional: each individual genome contains millions of variants, many of which exhibit complex correlations due to linkage disequilibrium. Moreover, evolutionary processes operate at multiple scales—from individual loci to genome-wide patterns—making it difficult to model these systems using simple statistical assumptions. AI models are well suited to this complexity because they can learn hierarchical representations that reflect both local genetic patterns and global population structure [29]. This complexity motivates the application of AI methods for uncovering population structure and ancestry relationships.

4.6.3 AI for Population Structure and Ancestry Inference

One of the core tasks in population genomics is the identification of population structure patterns of genetic similarity that reflect shared ancestry or historical gene flow. Classical approaches rely on dimensionality reduction or probabilistic clustering, which often require assumptions about population equilibrium or predefined numbers of ancestral groups.

AI-based methods, including unsupervised and semi-supervised learning, offer more flexible alternatives. Autoencoders and manifold learning techniques compress high-dimensional genotype data into low-dimensional latent representations that capture ancestry gradients and admixture patterns. These representations often align with known geographic or historical relationships while also revealing finer-scale substructure [30].

Deep learning models have also been applied to infer local ancestry along the genome, identifying chromosomal segments inherited from different ancestral populations. This has practical relevance for studying admixture in human populations and for correcting population stratification in downstream association analyses.

Practical Example: In global population studies, AI-driven ancestry inference has been used to resolve subtle subpopulation differences within continental groups, improving the resolution of demographic reconstructions without asserting sociopolitical definitions of ancestry.

Beyond structure, population genomics seeks to identify evolutionary forces acting on specific genomic regions.

4.6.4 Detecting Selection and Adaptive Signals Using AI

Identifying genomic regions under natural selection is central to understanding adaptation and evolutionary history. Selection can leave complex signatures, including extended haplotype structures, shifts in allele frequencies, and changes in linkage patterns. These signals are often confounded by demographic processes, making them difficult to disentangle using traditional summary statistics alone.

AI models trained on simulated or empirical genomic data can learn composite patterns indicative of selection without relying on single metrics. Supervised classifiers distinguish between neutral and selected regions by integrating multiple features, such as haplotype diversity, allele frequency spectra, and recombination rates [31]. More recently, deep learning approaches have demonstrated the ability to detect soft sweeps and polygenic adaptation, which involve subtle changes across many loci.

Biological Relevance: These methods enable researchers to identify candidate genes and pathways involved in adaptation to environmental factors such as climate, diet, or pathogen exposure, providing evolutionary context for modern phenotypic variation.

Population genomics also informs medical genomics by contextualizing disease-associated variants within evolutionary and demographic frameworks.

4.6.5 Population Genomics, Disease Risk, and Health Disparities

AI-driven population genomics plays an increasingly important role in understanding how genetic risk varies across populations. Disease-associated variants identified in one ancestry group may not have the same frequency or effect size in others due to differences in linkage structure and evolutionary history. AI models can integrate population-specific genetic patterns to improve variant interpretation and risk stratification in diverse cohorts [32].

From a public health perspective, population-aware AI models help highlight gaps in representation and reduce biases arising from Eurocentric genomic datasets. While these models support research into genetic architecture and risk prediction, they are not substitutes for clinical validation and should not be interpreted as providing population-level medical recommendations.

Practical example: In pharmacogenomics, AI models incorporating population genomic data assist in identifying variants affecting drug metabolism that differ across ancestries, supporting research into safer and more equitable therapeutic strategies.

As population genomics expands, ethical and methodological challenges become increasingly salient.

4.6.6 Challenges, Bias, and Responsible Use of AI

Despite their promise, AI applications in population genomics raise important concerns. Models trained on imbalanced datasets may reinforce existing disparities or misinterpret population structure as biological determinism. Furthermore, population labels are often proxies for complex social, cultural, and historical factors rather than discrete biological categories.

Responsible AI practice requires careful dataset curation, transparency in model assumptions, and avoidance of overinterpretation. Explainable AI techniques are increasingly used to clarify which genomic features drive model predictions, enabling biological validation and mitigating misuse [33]. Importantly, AI-based population genomic analyses are research tools and should not be conflated with ancestry-based clinical decision-making.

4.7 Summary

Artificial intelligence has become a foundational enabler across the full spectrum of modern genomic analysis, transforming how genomic data is generated, interpreted, and contextualized. Beginning at the earliest stages of data acquisition, AI-assisted base calling and genome assembly address long-standing challenges related to sequencing noise, repetitive regions, and large, complex genomes. By learning probabilistic mappings from raw signals to nucleotide sequences and refining classical graph-based assembly frameworks, AI improves the accuracy and completeness of genomic reconstructions while preserving biological fidelity.

Beyond sequence generation, AI-driven variant calling and prioritization enhance the detection and interpretation of genetic variation, particularly in difficult genomic contexts and non-coding regions. By integrating sequence features with functional annotations, evolutionary conservation, and phenotypic information, AI models provide probabilistic rankings that guide researchers toward biologically relevant variants. These approaches improve consistency across heterogeneous datasets and support discovery-oriented research in rare disease, cancer genomics, and functional genomics, while emphasizing the need for careful validation and interpretability.

AI-based modeling of gene expression and epigenomic regulation further deepens our understanding of genome function by capturing hierarchical, multi-scale regulatory interactions. Through the integration of DNA sequence, chromatin accessibility, histone modifications, methylation patterns, and three-dimensional genome organization, AI models offer scalable insights into transcriptional regulation across tissues, developmental stages, and disease states. While challenges related to bias, causal inference, and model transparency persist, these methods serve as powerful hypothesis-generating tools that complement experimental assays.

In the context of complex traits, AI-driven disease risk prediction and polygenic score modeling enable nonlinear analysis of variant interactions and multi-omic integration. These frameworks provide biologically informed perspectives on disease susceptibility across populations, supporting research in cardiovascular disease, cancer, neurodegeneration, and pharmacogenomics. Ethical considerations, population diversity, and responsible deployment remain central to ensuring that such models inform research rather than overextend into unvalidated clinical use.

AI also plays a critical role in microbiome and metagenomic analysis, where it addresses the scale, heterogeneity, and compositional complexity of microbial data. By enabling taxonomic classification, functional inference, host–microbe interaction modeling, and temporal ecosystem analysis, AI facilitates deeper biological insight into microbial communities while maintaining a research-driven, non-causal interpretative framework.

Finally, AI has expanded the analytical scope of population genomics by enabling scalable inference of ancestry, detection of selection, and population-aware interpretation of genetic variation. When integrated with classical population genetics principles, AI contributes valuable evolutionary insights while reinforcing the importance of transparency, diversity, and ethical responsibility.

Collectively, the applications discussed in this chapter illustrate that AI is not a replacement for biological theory or experimental validation, but a powerful complementary framework. When responsibly applied, AI strengthens every layer of genomic analysis—from raw data generation to evolutionary interpretation—advancing discovery, supporting translational research, and shaping the future of precision genomics.

References

1. Zubair, M., Khan, A. H., Bilal, S. F., & Li, J. (2025). Deep learning approaches for resolving genomic discrepancies in cancer: A systematic review and clinical perspective. *Briefings in Bioinformatics, 26*(6), bbaf541. https://doi.org/10.1093/bib/bbaf541
2. Athanasopoulou, K., Michalopoulou, V. I., Scorilas, A., & Adamopoulos, P. G. (2025). Integrating artificial intelligence in next-generation sequencing: Advances, challenges, and future directions. *Current Issues in Molecular Biology, 47*, 470. https://doi.org/10.3390/cimb47060470

3. Warburton, P. E., & Sebra, R. P. (2024). Long-read DNA sequencing: Recent advances and remaining challenges. *Annual Review of Genomics and Human Genetics, 26*, 79–108. https://doi.org/10.1146/annurev-genom-101722-103045
4. Luan, T., et al. (2025). MetaCompass. *Cell Genomics, 5*, 101186. https://doi.org/10.1016/j.crmeth.2025.101186
5. Vaser, R., Sović, I., & Holt, K. E. (2025). Autocycler: Long-read consensus assembly for bacterial genomes. *Bioinformatics, 41*, btaf474. https://doi.org/10.1093/bioinformatics/btaf474
6. Cannon, E. K. S., et al. (2025). Guidelines for gene and genome assembly nomenclature. *Genetics, 229*, iyaf006. https://doi.org/10.1093/genetics/iyaf006
7. Abdelwahab, O., et al. (2025). AI variant calling review. *Frontiers in Bioinformatics, 5*, 1574359. https://doi.org/10.3389/fbinf.2025.1574359
8. Zheng, Z., et al. (2025). Clair3-RNA. *Nature Communications, 16*, 67237. https://doi.org/10.1038/s41467-025-67237-6
9. Hsu, C. Y., Askar, S., Alshkarchy, S. S., Nayak, P. P., et al. (2025). AI-driven multi-omics integration in precision oncology: Bridging the data deluge to clinical decisions. *Clinical and Experimental Medicine, 26*, 29. https://doi.org/10.1007/s10238-025-01965-9
10. Lekadir, K., et al. (2025). FUTURE-AI: International consensus guideline for trustworthy artificial intelligence in healthcare. *BMJ, 388*, e081554. https://doi.org/10.1136/bmj-2024-081554
11. Barbadilla-Martínez, L., Klaassen, N., van Steensel, B., & de Ridder, J. (2025). Predicting gene expression from DNA sequence using deep learning models. *Nature Reviews Genetics, 26*, 666–680. https://doi.org/10.1038/s41576-025-00841-2
12. Avsec, Ž, et al. (2021). Effective gene expression prediction from sequence by integrating long-range interactions. *Nature Methods, 18*, 1196–1203. https://doi.org/10.1038/s41592-021-01252-x
13. Teji, B., et al. (2025). GT-GRN: A graph transformer framework for enhanced gene regulatory network inference. *Briefings in Bioinformatics, 26*, bbaf584. https://doi.org/10.1093/bib/bbaf584
14. Li, Y., et al. (2025). GRANet: A graph residual attention network for gene regulatory network inference. *Briefings in Bioinformatics, 26*, bbaf349. https://doi.org/10.1093/bib/bbaf349
15. Avsec, Ž, Latysheva, N., Kohli, P., et al. (2026). Advancing regulatory variant effect prediction with AlphaGenome. *Nature, 635*, 1206–1218. https://doi.org/10.1038/s41586-025-10014-0
16. Javed, N., et al. (2025). Multi-modal transformer. *Cell Genomics, 5*, 100762. https://doi.org/10.1016/j.xgen.2025.100762
17. Schuette, G., Lao, Z., & Zhang, B. (2025). ChromoGen: diffusion model predicts single-cell chromatin conformations. *Science Advances, 11*, eadr8265. https://doi.org/10.1126/sciadv.adr8265
18. Schuran, M., et al. (2025). A survey on deep learning for polygenic risk scores. *Briefings in Bioinformatics, 26*, bbaf373. https://doi.org/10.1093/bib/bbaf373
19. Team, A., et al. (2026). Multimodal deep learning approaches for improving polygenic prediction with imaging data. *Scientific Reports, 16*, 34153. https://doi.org/10.1038/s41598-025-34153-6
20. Kelemen, M., Zhao, J. H., & Inouye, M. (2025). Performance of deep-learning-based approaches to improve polygenic scores. *Nature Communications, 16*, 60056. https://doi.org/10.1038/s41467-025-60056-1
21. Cannoodt, R., et al. (2024). Graph convolutional networks for polygenic risk prediction across ancestries. *Nature Genetics, 56*, 2228–2237. https://doi.org/10.1038/s41588-024-01893-5
22. Fahed, A. C., et al. (2022). The potential of polygenic scores to improve cost and outcomes of clinical trials. *Nature Communications, 13*, 3164. https://doi.org/10.1038/s41467-022-30675-z

23. Xiao, L., et al. (2024). Exploring the frontier of microbiome biomarker discovery with artificial intelligence. *National Science Review, 11*, nwae325. https://doi.org/10.1093/nsr/nwae325
24. Karin, E. L., et al. (2025). Cutting-edge deep-learning based tools for metagenomic classification. *National Science Review, 12*, nwaf056. https://doi.org/10.1093/nsr/nwaf056
25. Sharma, S., et al. (2025). ML metagenomics. *mSystems, 10*, e01642-24. https://doi.org/10.1128/msystems.01642-24
26. Khera, A. V., et al. (2023). Performance of polygenic risk scores in screening, prediction, and risk stratification: Secondary analysis of Polygenic Score Catalog. *BMJ Medicine, 2*, e000554. https://doi.org/10.1136/bmjmed-2022-000554
27. Zhou, T., et al. (2025). AI microbiome. *Gut, 74*, 335946. https://doi.org/10.1136/gutjnl-2025-335946
28. Byrska-Bishop, M., et al. (2025). Whole-genome sequencing of 490,640 UK Biobank participants. *Nature, 643*, 211–219. https://doi.org/10.1038/s41586-025-09272-9
29. Stern, A., et al. (2023). ML evolutionary studies. *Proceedings of the National Academy of Sciences of the USA, 120*, e2205058119. https://doi.org/10.1073/pnas.2205058119
30. Silvestro, D., & Salamin, N. (2024). PhyloRNN: Deep learning for phylogenetic inference and population structure. *Systematic Biology, 73*, syae029. https://doi.org/10.1093/sysbio/syae029
31. Arellano Davín, A. A., et al. (2025). A geological timescale for bacterial evolution and oxygen adaptation. *Science, 388*, 102–115. https://doi.org/10.1126/science.adp1853
32. Korfmann, K., et al. (2023). Deep learning in population genetics. *Genome Biology and Evolution, 15*, evad008. https://doi.org/10.1093/gbe/evad008
33. Mathieson, S., et al. (2025). Interpreting convolutional neural networks in population genetics. bioRxiv. https://doi.org/10.64898/2025.12.04.692404

5 AI in Precision Medicine and Drug Discovery

The historical paradigm of medical treatment has long been dictated by an empirical, generalized methodology. This traditional "one-size-fits-all" approach, while instrumental in combating broad population-level health crises, inherently relies on trial and error. Consequently, this methodology leaves significant sub-populations of patients deriving little to no therapeutic benefit from standard interventions, or worse, suffering from severe, sometimes fatal, adverse drug reactions [1]. The accelerated growth of large research cohorts, coupled with the digitization of health-related data, is currently fueling a fundamental transformation away from empirical medicine and toward precision medicine [1]. Precision medicine leverages sophisticated computational tools to identify unique patient phenotypes, allowing clinicians to tailor preventative measures and personalized health management protocols based on an individual's distinct genetic variations, physiological indicators, and lifestyle factors [1].

Concurrently, the global pharmaceutical industry is navigating a severe productivity and economic crisis. The financial burden of developing a novel therapeutic and successfully navigating it through regulatory approval has escalated dramatically. Recent estimates place the total life cycle cost of bringing a single drug product to market at \$2.56 billion, with an additional \$300 million typically allocated for post-approval marketing surveillance, bringing the comprehensive cost to approximately \$2.87 billion [2]. This staggering financial investment is further compounded by rising drug attrition rates, representing massive losses for pharmaceutical entities [2].

K. Shaikh et al., *Artificial Intelligence in Genomics*, Synthesis Lectures on Engineering, Science, and Technology, https://doi.org/10.1007/978-3-032-25486-3_5

The convergence of artificial intelligence (AI) with precision medicine and drug discovery provides unprecedented opportunities to mitigate these systemic inefficiencies [1]. The widespread adoption of AI in pharmaceutical pipelines, catalyzed in the 2010s by advances in big data and deep learning architectures, has transformed AI into an indispensable asset across the entire drug development lifecycle, from initial target identification to the intricate design of clinical trials [3]. By integrating human phenomics, genomic data, and vast chemical libraries, machine learning models possess the capacity to predict disease risk, stratify patient cohorts, elucidate biological mechanisms, and generate entirely novel therapeutic compounds with optimized pharmacological profiles [4].

This chapter provides a rigorous, exhaustive technical exploration of the state-of-the-art AI applications in precision medicine and drug discovery. The analysis details the underlying mathematical concepts, reviews the dominant computational frameworks, explores large-scale multi-omics integration strategies, and provides practical, reproducible code implementations for core tasks including patient stratification, biomarker discovery, molecular property prediction, and generative chemistry.

5.1 Foundational Infrastructure: Software Frameworks and Data Ecosystems

The successful application of artificial intelligence in the life sciences is entirely contingent upon the availability of high-throughput, high-quality data and the computational frameworks capable of processing them at scale. Healthcare data is expanding at an incomprehensible rate, with projections indicating growth to 2.3 zettabytes of information [2]. However, the sheer volume of data is only one facet of the challenge; the structural nature of the data is equally problematic. An estimated eighty percent of modern healthcare data exists in completely unstructured and predominantly unlabeled formats, encompassing physician notes, medical imaging, and continuous sensor data [2]. Extracting actionable biological insights from this chaotic data landscape requires advanced natural language processing (NLP), computer vision, and highly scalable machine learning infrastructure.

5.1.1 Dominant Computational and Machine Learning Frameworks

The modern bioinformatics and cheminformatics landscape relies on a stack of highly optimized, open-source programming libraries that bridge the gap between theoretical mathematics and scalable, production-ready deployment.

The foundational layer of deep learning research and production is currently dominated by PyTorch and TensorFlow. PyTorch, originating from Meta's AI Research lab, has surged to prominence and become the definitive favorite in academic research, particularly within geometric deep learning and generative chemistry [5]. Its primary advantage lies in its dynamic computational graph, which allows for intuitive, flexible model building, a critical feature when designing complex, non-Euclidean architectures such as Graph Neural Networks (GNNs) for molecular modeling [5]. Conversely, TensorFlow, backed by Google, remains a production powerhouse. It excels in deploying models at scale through tools like TensorFlow Extended (TFX) and TensorFlow Lite for edge AI devices, supporting highly distributed training across CPUs, GPUs, and Tensor Processing Units (TPUs) [5].

For traditional machine learning algorithms, Scikit-learn remains the definitive industry standard. It provides highly optimized, user-friendly implementations of classical algorithms necessary for clinical data analysis, including Random Forests, Support Vector Machines (SVMs), and k-Means clustering, making it essential for tasks like biomarker discovery and initial exploratory data analysis [5]. XGBoost, an optimized gradient-boosting framework, is frequently deployed alongside Scikit-learn to handle structured tabular data, excelling in tasks such as entity resolution and predicting clinical outcomes from electronic health records [6].

When processing multi-omics datasets that exceed the memory capacity of single machines, frameworks designed for big data become necessary. Apache Spark MLlib provides distributed, highly scalable machine learning across clusters, integrating seamlessly with the broader big data ecosystem for both batch and real-time processing [5]. For hardware-accelerated pipelines, NVIDIA's RAPIDS.AI suite is heavily utilized to execute entire machine learning workflows from data loading and transformation to training and inference directly on GPUs, drastically reducing processing times for massive genomic cohorts [7].

Within the highly specialized domains of drug discovery and cheminformatics, domain-specific software solutions augment these general-purpose frameworks. Chemical Computing Group's Molecular Operating Environment (MOE) offers a comprehensive platform integrating molecular modeling, structure-based drug design, molecular docking, and Quantitative Structure–Activity Relationship (QSAR) modeling [8]. Open-source alternatives like Data Warrior provide robust chemical intelligence, allowing researchers to compute molecular descriptors and develop QSAR models using supervised machine learning [8]. Furthermore, libraries such as DeepChem and Torch Drug operate as high-level wrappers built atop PyTorch and TensorFlow, specifically engineered to democratize deep learning for molecular sciences by providing pre-trained models, specialized data loaders, and domain-specific featurizers [9].

5.1.2 Curated Public Datasets for Precision Medicine

The democratization of artificial intelligence in the biomedical sciences has been heavily driven by the availability of large-scale, publicly accessible datasets. These repositories provide the raw material necessary to train robust AI models, enabling a scale of analysis that was previously unattainable for individual research laboratories [10]. Table 5.1 details the primary datasets utilized for training AI models in precision medicine, biomarker discovery, and therapeutic design.

5.1.3 Practical Implementation: Accessing the Therapeutics Data Commons (TDC)

Data wrangling, the process of cleaning, structuring, and harmonizing disparate data sources, often consumes most of a data scientist's time. The Therapeutics Data Commons (TDC) was developed specifically to eliminate this bottleneck in drug discovery. TDC provides an ecosystem of AI-ready datasets, programmatic oracles, and curated benchmarks that map directly to the drug discovery and development pipeline, including target discovery, activity modeling, efficacy, safety, and manufacturing [13]. TDC organizes its resources through a strict three-tiered hierarchical structure designed for programmatic access [13]. The first tier defines the overarching machine learning problem: single-instance prediction, multi-instance prediction, or generative modeling. The second tier defines the specific therapeutic learning task, such as predicting ADME (Absorption, Distribution, Metabolism, and Excretion) properties, drug-target interactions (DTI), or protein–protein interactions (PPI). The third tier specifies the exact experimental dataset, such as the Caco-2 cell permeability assay dataset [13].

Table 5.1 Details of primary datasets for training AI models in precision medicine, biomarker discovery, and therapeutic design

Dataset repository	Primary data modality	Key focus and research application
The Cancer Genome Atlas (TCGA)	Multi-omics (genomic, transcriptomic, proteomic, Epigenomic)	Comprehensive molecular characterization of over 33 human cancer types, serving as the foundational dataset for cancer patient stratification and survival analysis [10]
Cancer Cell Line Encyclopedia (CCLE)	Omics and phenotypic	Extensive genomic characterization coupled with high-throughput drug sensitivity screening data for human cancer cell lines, vital for predicting drug response and resistance mechanisms [10]
Library of Integrated Network-Based Cellular Signatures (LINCS)	Omics and phenotypic	Catalogs changes in cellular transcriptomic signatures in response to a vast array of chemical and genetic perturbations, enabling the mapping of drug mechanisms [10]
ProteomeXchange	Proteomics	A globally coordinated repository for mass spectrometry-based proteomics data, facilitating the discovery of protein-level biomarkers [10]
CellXGene and HTAN	Single-cell omics and spatial transcriptomics	Provides highly granular single-cell transcriptomic data and high-dimensional spatial atlases of human tumors, allowing for the mapping of the tumor microenvironment [10]
1000 Genomes Project	Whole-genome sequencing (WGS)	The largest fully open resource of WGS data, cataloging common human genetic variation to serve as a healthy baseline for comparative genetic studies [11]
All of Us Research Program	WGS, EHR, wearable sensors	A massive initiative collecting longitudinal data from over one million diverse participants to advance precision medicine, linking genetics directly to clinical health records [11]
TDC (Therapeutics Data Commons)	AI/ML-ready curated datasets	A multimodal foundation providing systematically curated datasets and benchmarks across all stages of therapeutic discovery, from target identification to molecule generation [12]
BioSNAP and OGB	Biomedical networks/ graphs	Collections of diverse biomedical networks, including protein–protein interactions and drug-drug interactions, designed specifically for training graph neural networks [12]

The following Python implementation demonstrates how to programmatically access TDC, retrieve an ADME dataset, and prepare it for machine learning using a domain-specific data splitting strategy.

```
# The PyTDC library is designed for minimal dependency overhead
# Installation command: pip install PyTDC --upgrade
from tdc.single_pred import ADME
from tdc.utils import retrieve_dataset_names
import pandas as pd

# 1. Dataset Exploration
# Retrieve and display all available datasets belonging to the ADME therapeutic task
adme_datasets = retrieve_dataset_names('ADME')
print(f"Available ADME predictive datasets: {adme_datasets[:5]}...")

# 2. Data Retrieval
# Instantiate the Caco-2 (Cell Permeability) dataset class from the Wang et al. study
# This dataset contains SMILES strings of molecules and their experimental permeability
values
data = ADME(name='Caco2_Wang')

# Extract the entire dataset into a standard Pandas DataFrame for initial exploratory
analysis
df_caco2 = data.get_data(format='df')
print("\nDataset Sample:")
print(df_caco2.head())

# 3. Methodological Data Splitting
# In computational chemistry, standard random splitting often leads to severe model
overestimation.
# Structurally identical molecules with only minor functional group variations may appear
in
# both the training and testing sets, allowing the model to memorize rather than learn.
# TDC provides the 'scaffold' split method, which clusters molecules by their Bemis-
Murcko
# molecular scaffold, ensuring that the test set contains entirely novel chemical core
structures.
# This accurately simulates the real-world challenge of generalizing to unseen regions of
chemical space.
data_splits = data.get_split(method='scaffold', seed=42, frac=[0.7, 0.1, 0.2])

train_df = data_splits['train']
val_df = data_splits['valid']
test_df = data_splits['test']

print(f"\nTraining set size: {len(train_df)} molecules")
print(f"Test set size: {len(test_df)} molecules")

# 4. Model Evaluation Preparation
# TDC also provides standardized evaluators to ensure models are benchmarked fairly
from tdc import Evaluator
# For regression tasks like Caco-2 permeability, Mean Absolute Error (MAE) is commonly
used
evaluator = Evaluator(name='MAE')
# Usage post-prediction: mae_score = evaluator(y_true, y_pred)
```

The output of Python code is as follows:

```
Found local copy...
Loading...
Done!
Available ADME predictive datasets: ['lipophilicity_astrazeneca',
'solubility_aqsoldb', 'hydrationfreeenergy_freesolv', 'caco2_wang',
'pampa_ncats']...

Dataset Sample:
                                                Drug_ID  \
0                                         (-)-epicatechin
1   (2E,4Z,8Z)-N-isobutyldodeca-2,4,10-triene-8 -y...
2                                                codeine
3                                             creatinine
4                                                danazol

                                                   Drug         Y
0              Oc1cc(O)c2c(c1)OC(c1ccc(O)c(O)c1)C(O)C2 -6.220000
1                    C/C=C\C#CCC/C=C\C=C\C(=O)NCC(C)C -3.860000
2   COc1ccc2c3c1O[C@H]1[C@@H](O)C=C[C@H]4[C@@H](C2... -4.090000
3                                    CN1CC(=O)NC1=N -5.935409
4   C#C[C@]1(O)CC[C@H]2[C@@H]3CCC4=Cc5oncc5C[C@]4(... -4.840000
100%|██████████| 910/910 [00:00<00:00, 1793.85it/s]

Training set size: 637 molecules
Test set size: 182 molecules
```

By providing these programmatic hooks, TDC enables researchers to rapidly prototype machine learning architectures without spending weeks harmonizing chemical formats or debating the statistical validity of their cross-validation folds.

5.2 Artificial Intelligence in Precision Medicine: Deciphering Disease Heterogeneity

The foundational premise of precision medicine is the acknowledgment of profound disease heterogeneity. Patients diagnosed with ostensibly identical conditions such as breast invasive carcinoma or lung adenocarcinoma frequently exhibit vastly different molecular drivers, dictating highly divergent prognoses and varying sensitivities to targeted kinase inhibitors or immunotherapies [14]. Artificial intelligence models serve as the analytical engine to decode this complexity, utilizing multi-omics data to stratify patients, identify causative biomarkers, and construct patient-specific clinical models.

5.2.1 Multi-omics Data Integration and Network Biology

Biological systems are not governed by a single molecular layer; they operate through a highly complex, interconnected cascade of genomics, epigenomics, transcriptomics (RNA-seq), proteomics, and metabolomics. Relying on a single modality provides a limited snapshot view of the cellular state. The integration of multi-omics data provides a holistic understanding of disease mechanisms, but it introduces monumental analytical challenges due to varying data scales, high noise-to-signal ratios, and the presence of missing or incomplete data across patient cohorts [1].

To extract unified insights from these disparate layers, computational biologists employ advanced integration frameworks. Unsupervised multi-omics network methods are generally categorized into distinct statistical approaches. Matrix factorization-based networks, such as Collective Matrix Factorization (CMF) or Network-Based Stratification (NBS), attempt to decompose multiple omics matrices simultaneously to discover shared latent features that explain patient variance [15]. Network-based methods, alternatively, leverage graph topologies. A highly utilized algorithm in this domain is Similarity Network Fusion (SNF). SNF constructs independent patient-to-patient similarity networks for each distinct omics layer (e.g., one network for mRNA expression, another for DNA methylation). Through an iterative message-passing procedure, SNF fuses these separate networks into a single, comprehensive patient similarity graph, capturing both the strong commonalities across all layers and the unique signals present in individual modalities [16].

Emerging state-of-the-art architecture utilizes deep learning to achieve this fusion. Frameworks such as Integrate Any Omics (IntegrAO) leverage Graph Neural Networks (GNNs) applied to patient similarity graphs. IntegrAO is particularly notable for its ability to handle missing modalities, a ubiquitous challenge where some patients in a clinical trial may have RNA-seq data but lack proteomics. The model generates unified, genome-wide patient embeddings that can be subsequently utilized for unsupervised stratification or supervised clinical classification [17]. Alternatively, tools like PathIntegrate shift the integration paradigm from the molecular level to the biological systems level. By employing single-sample pathway analysis, PathIntegrate transforms raw multi-omics datasets into interpretable pathway-level representations, allowing predictive models to rank the exact biological pathways contributing to specific clinical outcomes, such as the immune dysregulation observed in COVID-19 or COPD [18].

5.2.2 Unsupervised Learning for Patient Stratification

Patient stratification aims to partition a highly heterogeneous patient cohort into distinct, clinically meaningful subgroups based on their underlying molecular profiles. Because the true biological subtypes of a disease are often unknown a priori, this is formulated as an unsupervised machine learning problem [19]. High-throughput sequencing technologies, such as RNA-seq, provide quantitative expression levels for tens of thousands of genes simultaneously. However, clustering algorithms operating in this high-dimensional space fall victim to the "curse of dimensionality," wherein the mathematical distance between any two data points becomes statistically uniform, rendering distance-based clustering algorithms ineffective [20].

To circumvent this, stratification pipelines typically execute a rigorous sequence of dimensionality reduction followed by robust clustering [20]. Techniques such as Principal Component Analysis (PCA) or deep learning autoencoders are utilized to project the high-dimensional omics data into a dense, lower-dimensional latent space [20]. Following projection, clustering algorithms such as K-Means, Spectral Clustering utilizing Radial Basis Function (RBF) kernels, or Agglomerative Hierarchical Clustering are applied to isolate the patient subpopulations [20].

- **Practical Implementation: TCGA RNA-Seq Clustering Pipeline**: The following Python implementation demonstrates a highly structured unsupervised learning pipeline designed to stratify a simulated patient cohort based on RNA-seq gene expression data. The script utilizes the Scikit-learn library to execute data standardization, dimensionality reduction via PCA, and K-Means clustering, while utilizing the Silhouette score to mathematically determine the optimal number of disease subtypes [21].

```
import numpy as np
import pandas as pd
from sklearn.preprocessing import RobustScaler
from sklearn.decomposition import PCA
from sklearn.cluster import KMeans
from sklearn.metrics import silhouette_score
import matplotlib.pyplot as plt
import seaborn as sns

# 1. Ingest Gene Expression Data (Simulating a TCGA RNA-Seq Cohort)
# Structure: Rows represent distinct patient samples; Columns represent specific gene
expression counts.
# Simulating a cohort of 800 patients across 5,000 highly variable genes.
np.random.seed(42)
X_rna_seq = np.random.lognormal(mean=2.0, sigma=1.0, size=(800, 5000))

# 2. Advanced Data Preprocessing
# RNA-seq read counts are inherently skewed and prone to extreme biological outliers.
# Standard scaling (zero mean, unit variance) is highly sensitive to these outliers.
# The RobustScaler utilizes median and interquartile range (IQR) to standardize the data,
# ensuring that extreme gene expression values do not disproportionately warp the
clustering manifold.
scaler = RobustScaler()
X_scaled = scaler.fit_transform(X_rna_seq)

# 3. Dimensionality Reduction (PCA)
# We project the 5,000-dimensional gene space into a manageable 50-dimensional principal
component space.
# This isolates the primary axes of biological variance while filtering out transcriptional
noise.
pca = PCA(n_components=50, random_state=42)
X_pca = pca.fit_transform(X_scaled)

explained_variance = np.sum(pca.explained_variance_ratio_) * 100
print(f"Cumulative variance captured by top 50 PCs: {explained_variance:.2f}%")

# 4. Hyperparameter Optimization: Determining the Optimal Number of Clusters (k)
# The true number of cancer subtypes is unknown. We iteratively evaluate k from 2 to 7.
# The Silhouette Score evaluates cluster cohesion (how close a patient is to their own
cluster) versus cluster separation (how far a patient is from neighboring clusters).
best_k = 2
max_silhouette = -1.0
silhouette_scores = []

for k in range(2, 8):
    # K-Means algorithm partitions the variance into k distinct centroids
    kmeans_eval = KMeans(n_clusters=k, init='k-means++', n_init=10, random_state=42)
    cluster_assignments = kmeans_eval.fit_predict(X_pca)
```

```
        score = silhouette_score(X_pca, cluster_assignments)
        silhouette_scores.append(score)
        print(f"Evaluated k={k}: Silhouette Score = {score:.4f}")

        if score > max_silhouette:
            max_silhouette = score
            best_k = k

print(f"\nOptimal patient stratification achieved at k={best_k} distinct subtypes.")

# 5. Final Model Instantiation and Patient Assignment
final_model = KMeans(n_clusters=best_k, init='k-means++', n_init=20, random_state=42)
final_patient_strata = final_model.fit_predict(X_pca)

# The 'final_patient_strata' array now acts as a new, computationally derived clinical label.
# In a full clinical pipeline, these clusters are subsequently analyzed using Kaplan-Meier
survival curves to determine if the discovered subtypes experience statistically significant
differences in overall survival or treatment responsiveness.
```

The output of Python code for is as follows:

```
Cumulative variance captured by top 50 PCs: 11.46%
Evaluated k=2: Silhouette Score = 0.0166
Evaluated k=3: Silhouette Score = 0.0129
Evaluated k=4: Silhouette Score = 0.0152
Evaluated k=5: Silhouette Score = 0.0123
Evaluated k=6: Silhouette Score = -0.0010
Evaluated k=7: Silhouette Score = -0.0010

Optimal patient stratification achieved at k=2 distinct
subtypes.
```

Advanced implementations of this pipeline frequently substitute the standard Euclidean distance metric with variance-weighted cosine distance. Because gene expression pathways activate in proportional cascades rather than absolute magnitudes, the angle of the multi-dimensional expression vector is often far more biologically informative than its length. Implementing cosine distance within Agglomerative clustering frameworks has been empirically shown to drastically improve silhouette scores from roughly 12.5% to over 74% in complex RNA-seq datasets [20].

5.2.3 Supervised Biomarker Identification and Feature Selection

Once distinct patient strata are identified, or when analyzing predefined clinical cohorts (e.g., responders versus non-responders to a specific immunotherapy), the analytical objective shifts from unsupervised clustering to supervised feature selection. The goal is to identify the precise biomarkers specific genes, proteins, or clinical measurements that are driving the disease state or therapeutic resistance [22]. Identifying a robust biomarker signature is essential for developing rapid diagnostic assays and understanding fundamental biological mechanisms. For instance, applying single-cell transcriptomics to immune dysregulation has revealed that the differential expression of genes such as CD3, CD14, CD16, FOSB, S100A12, and TCRγδ can accurately distinguish between mild and severe clinical presentations of sepsis and COVID-19 [23].

Biomarker identification is mathematically formulated as a high-dimensional feature selection problem. Comprehensive, production-grade pipelines, such as the Clinical Biomarker Detection framework, implement a rigorous, multi-stage filtration process prior to model training [22]. These pipelines execute constancy checks to remove low-variance genes, apply correlation filters (e.g., Spearman's Rank Order Correlation for continuous variables and Corrected Cramer's V for categorical features) to remove highly redundant multi-collinear features, and utilize sophisticated Imbalance Resampling techniques like SMOTE (Synthetic Minority Over-sampling Technique) to ensure that minority patient classes do not induce algorithmic bias [22].

Following preprocessing, tree-based ensemble algorithms, most notably the Random Forest classifier, are the tools of choice for extracting feature importance [24]. Unlike generalized linear models, Random Forests inherently capture complex, nonlinear, and non-additive biological interactions without requiring the explicit mathematical definition of interaction terms [25].

- **Practical Implementation: Random Forest Biomarker Extraction**

The algorithm constructs a multitude of distinct decision trees utilizing bootstrapped samples of the patient dataset. At each node split, the algorithm evaluates a random subset of genes and selects the one that maximizes the decrease in Gini impurity, 3a measure of how well a specific gene separates the healthy patients from the diseased patients. By averaging this impurity decrease across all trees in the forest, the model assigns a highly robust quantitative "Feature Importance" score to every gene [26].

```
from sklearn.ensemble import RandomForestClassifier
from sklearn.model_selection import train_test_split
from sklearn.metrics import classification_report, roc_auc_score

# 1. Formulate the Supervised Learning Dataset
# X_omics represents the scaled gene expression matrix.
# y_clinical represents the binary clinical outcome (e.g., 0 = Responder, 1 = Non-
Responder).
X_omics = X_scaled # Utilizing the RobustScaled data from the previous pipeline
y_clinical = np.random.randint(0, 2, size=800)

# Generate simulated gene nomenclature for interpretability in the final output
gene_identifiers = np.array([f'Gene_{i}' for i in range(X_omics.shape[1])])

# 2. Stratified Train-Test Split
# A 70-30 split is executed. The 'stratify' parameter ensures that the ratio of
Responders
# to Non-Responders is perfectly maintained across both training and testing
datasets.
X_train, X_test, y_train, y_test = train_test_split(
    X_omics, y_clinical, test_size=0.3, random_state=42, stratify=y_clinical
)

# 3. Model Initialization and Training
# max_features='sqrt' restricts each node split to evaluate only the square root of
total genes.
# This forces the trees to be highly diverse, preventing a few dominant genes from
masking
# secondary, subtle biomarker interactions.
rf_biomarker_model = RandomForestClassifier(
    n_estimators=1000,          # A large forest ensures statistical stability
    max_features='sqrt',        # Standard heuristic for classification trees
    class_weight='balanced',    # Automatically penalizes errors in the minority
patient class
    n_jobs=-1,                  # Utilize all available CPU cores for parallel processing
    random_state=42
)

print("Training Random Forest ensemble for feature extraction...")
rf_biomarker_model.fit(X_train, y_train)
```

```
# 4. Clinical Predictive Performance
# Before trusting the biomarkers, we must verify the model's predictive validity.
y_prob = rf_biomarker_model.predict_proba(X_test)[:, 1]
roc_auc = roc_auc_score(y_test, y_prob)
print(f"Model Predictive Power (ROC-AUC): {roc_auc:.3f}")

# 5. Extracting the Biomarker Signature
# We extract the Gini importance scores, sort them, and map them to their specific
gene identifiers.
gini_importances = rf_biomarker_model.feature_importances_
sorted_indices = np.argsort(gini_importances)[::-1] # Sort in descending order

print("\nTop 10 Biomarker Candidates (Drivers of Clinical Phenotype):")
print("-" * 65)
print(f"{'Rank':<5} | {'Ensembl Gene ID':<18} | {'Gini Importance Score':<25}")
print("-" * 65)

for i in range(10):
    rank = i + 1
    gene = gene_identifiers[sorted_indices[i]]
    score = gini_importances[sorted_indices[i]]
    print(f"{rank:<5} | {gene:<18} | {score:.6f}")
```

The output of random forest biomarker extraction is as follows:

```
Training Random Forest ensemble for feature
extraction...
Model Predictive Power (ROC-AUC): 0.517

Top 10 Biomarker Candidates (Drivers of Clinical
Phenotype):
-----------------------------------------------------------------
Rank  | Ensembl Gene ID    | Gini Importance Score
-----------------------------------------------------------------
1     | Gene_3577          | 0.001139
2     | Gene_682           | 0.001025
3     | Gene_2593          | 0.000954
4     | Gene_1750          | 0.000924
5     | Gene_2926          | 0.000912
6     | Gene_2477          | 0.000905
7     | Gene_4920          | 0.000890
8     | Gene_4398          | 0.000886
9     | Gene_1994          | 0.000879
10    | Gene_2869          | 0.000842
```

- **Dimensionality Reduction (PCA)**: Reducing 5000 gene dimensions to 50 principal components retains only 11.46% of total variance, which may risk losing important information. Higher variance retention is often preferred for robust data representation, though a lower percentage might suffice if it captures key biological signals.
- **Optimal Number of Clusters (k) Using Silhouette Score**: Silhouette Scores for k = 2 to k = 7 is all low, with k = 2 having the highest at 0.0166. Low or negative scores indicate weak separation between clusters, suggesting few distinct subtypes in the simulated dataset after PCA.

Overall, the pipeline completed scaling, dimensionality reduction, and clustering, but results show limited evidence of clear patient subtypes. Further feature engineering or different analytical methods may be needed. In highly correlated biological pathways, the standard Gini importance metric can occasionally dilute the signal across a family of co-expressed genes. To refining this, advanced specialized libraries such as pyRforest implement rank-based permutation methods. By systematically shuffling the values of individual genes and observing the resultant drop in model accuracy, researchers can derive robust, statistically rigorous P-values and SHapley Additive exPlanations (SHAP) values for individual biomarkers, allowing for direct downstream integration into Gene Ontology (GO) and pathway enrichment analyses [25].

5.2.4 Clinical Modeling and Survival Analysis

Connecting these omics-derived strata and biomarkers to longitudinal patient outcomes requires rigorous statistical modeling, primarily executed through survival analysis [27]. The industry standard for decades has been the Cox Proportional Hazards model. However, this mathematical model relies on a rigid assumption: that the effect of a predictor (e.g., the expression level of a biomarker) remains constant and proportional over time across all individuals [27]. In modern, diverse, multi-center precision medicine studies, this assumption is frequently violated. Baseline survival probabilities fluctuate drastically depending on the specific hospital, study center, regional demographics, or disparate local care protocols [27]. When the proportional hazards assumption fails, the standard Cox model generates highly misleading hazard ratios.

To rectify this mathematical violation, computational biologists deploy the stratified Cox model. This advanced formulation extends the standard model by allowing the underlying baseline hazard function, denoted as $h_0(t)$, to vary entirely independently across predefined strata (e.g., biological sexes or specific clinical trial sites). Crucially, while the baseline risks fluctuate, the model simultaneously estimates a single, common coefficient vector β for primary biological covariates (the biomarkers) across all groups. Let the hazard for patient I situated within stratum s be defined as:

$$h_{is}(t) = h_{os}(t)\exp\left(\beta^T X_{is}\right)$$

where X_{is} represents the vector of omics features. This sophisticated mathematical adjustment acts as a powerful statistical filter, ensuring that the algorithm successfully isolates the true biological signal of the biomarker from the systemic, confounding noise of disparate clinical environments [27].

5.3 Artificial Intelligence in Drug Discovery: From Biomarkers to Therapeutics

If precision medicine is the science of identifying the exact biological target, modern drug discovery is the engineering required to generate the precise chemical key to manipulate it. The application of deep learning to chemistry has initiated an era where computational systems can parse billions of compounds, accurately predict their highly complex physicochemical properties, and design entirely novel molecular structures de novo [3].

5.3.1 Target Identification and Drug-Target Interactions (DTI)

The bridge between a patient's genomic profile and therapeutic intervention relies on predicting how specific chemical compounds interact with targeted proteins. Drug-Target Interaction (DTI) prediction has evolved significantly. Traditional methods relied on basic statistical matching, but state-of-the-art architecture now constructs heterogeneous graphs integrating known drug structures, protein targets, and historical interaction data. Models such as DTI-HETA utilize Graph Attention (GAT) layers to learn deep node embeddings, achieving unprecedented accuracy in predicting binding affinities [28]. Simultaneously, simpler Multilayer Perceptron (MLP) architecture remains highly viable for rapid, initial exploratory analyses and target prioritization [28].

Furthermore, AI models are now capable of executing zero-shot predictions for therapeutic use. Systems such as TxGNN, powered by geometric deep learning and comprehensive knowledge graphs, allow researchers to query diseases and retrieve ranked lists of potential therapeutics based entirely on graph topology and meta-paths. For example, when querying Schizophrenia through the TxGNN framework, the model confidently assigns predictive scores of 1.000 to drugs like Mesoridazine, Iloperidone, and Chlorpromazine, leveraging complex meta-paths (e.g., "disease-gene/protein-disease-drug") to explain the biological rationale behind its predictions [13]. This capability is revolutionizing drug repurposing efforts by passing years of initial screening [13].

5.3.2 Cheminformatics and Molecular Representations

Machine learning algorithms are fundamentally mathematical constructs; they require numerical vectors, matrices, or tensors as input. Converting organic chemistry into a machine-readable format is the foundational challenge of cheminformatics. The prevailing standard is the Simplified Molecular-Input Line-Entry System (SMILES), which encodes the three-dimensional topology of a molecule into a dense, linear ASCII string [29].

While SMILES strings are highly efficient for database storage, they are linear representations of inherently non-linear entities. To capture true chemical reality, a molecule must be represented as an undirected graph $G = (V, E)$, where individual atoms constitute the node set V and the chemical bonds between them constitute the edge set E [30]. Modern AI pipelines rely extensively on the open-source library **RDKit**. RDKit acts as the translator, parsing SMILES strings, constructing navigable molecular graphs, and computing vital, localized chemical descriptors (such as partial charges, pharmacophores, and hybridization states) required for neural network input [29].

- **Practical Implementation: Molecular Processing with RDKit**

```
from rdkit import Chem
from rdkit.Chem import Descriptors
from rdkit.Chem.Draw import MolsToGridImage
import pandas as pd
from IPython.display import display, Image # Import display and Image from
IPython.display

# 1. Chemical Definition via SMILES
# Defining a small library of well-known therapeutic compounds
smiles_database = {
  "Aspirin": "CC(=O)Oc1ccccc1C(=O)O",
  "Caffeine": "CN1C=NC2=C1C(=O)N(C(=O)N2C)C",
  "Testosterone": "CC12CCC3C(C1CCC2O)CCC4=CC(=O)CCC34C"
}

# 2. Graph Construction
# RDKit converts the linear string into a computable molecular object
molecules = []
records = []

for name, smi in smiles_database.items():
  mol = Chem.MolFromSmiles(smi)
  if mol is not None:
    molecules.append(mol)
```

```
    # 3. Computing Critical Physicochemical Descriptors
        # Molecular Weight impacts formulation; LogP determines  lipophilicity and
        membrane permeability.
    mw = Descriptors.MolWt(mol)
    logp = Descriptors.MolLogP(mol)

    records.append({"Compound": name, "SMILES": smi, "MolWt": mw, "LogP": logp})

# Display the computed descriptors
df_properties = pd.DataFrame(records)
print(df_properties.to_string(index=False))

# 4. Visualization
# Generating 2D structural representations is critical for chemist interpretation.
# MolsToGridImage generates a standardized graphical grid of the parsed compounds.
img = MolsToGridImage(molecules, legends=list(smiles_database.keys()),
                molsPerRow=3, subImgSize=(300, 300))
# To display the RDKit image in Colab, use IPython.display.display
display(img)
```

The output of Molecular Processing with RDKit is as follows:

```
Compound                                    SMILES   MolWt    LogP
     Aspirin              CC(=O)Oc1ccccc1C(=O)O 180.159  1.3101
    Caffeine          CN1C=NC2=C1C(=O)N(C(=O)N2C)C 194.194 -1.0293
Testosterone CC12CCC3C(C1CCC2O)CCC4=CC(=O)CCC34C 288.431  3.8792
```

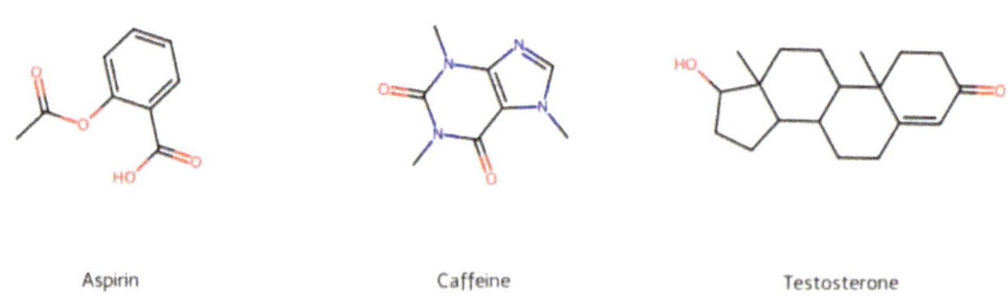

5.3.3 Graph Neural Networks for Molecular Property Prediction

Before a generated molecule can be considered a viable drug candidate, it must satisfy a highly stringent array of ADMET criteria. Predicting these properties computationally is referred to as Quantitative Structure–Activity Relationship (QSAR) modeling. Traditional QSAR methodologies required scientists to manually engineer molecular fingerprints (such as ECFP or MACCS keys) and feed them into standard classifiers [28].

State-of-the-art predictive modeling has abandoned manual feature engineering in favor of Graph Neural Networks (GNNs), which operate directly on the raw molecular graph [30]. GNNs utilize a powerful "Message Passing" paradigm. Within architectures like the Graph Convolutional Network (GCN) or the Graph Isomorphism Network (GIN), each individual atom (node) iteratively updates its hidden state representation by aggregating and transforming information from its immediate bonded neighbors [31]. After $\boldsymbol{k}$ message-passing layers, each node possesses a dense feature vector that perfectly encodes its $\boldsymbol{k}$-hop local chemical environment. A global pooling operation (such as sum or mean pooling) is then executed to aggregate these atom-level embeddings into a single, fixed-length graph-level vector, which is subsequently passed through fully connected neural layers for the final property regression or classification [31].

- **Practical Implementation: Property Prediction Using PyTorch Geometric (PyG)**

The PyTorch Geometric (PyG) library provides highly optimized, modular implementations of these complex message-passing layers [30]. The following implementation details the architecture and the foundational training loop for a GCN designed to predict a continuous chemical property, such as hydration free energy or aqueous solubility [30].

```
import torch
import torch.nn.functional as F
from torch.nn import Linear
from torch_geometric.nn import GCNConv, global_mean_pool
from torch_geometric.data import Data
from torch_geometric.loader import DataLoader

# 1. Defining the Graph Neural Network Architecture
class MolecularGCN(torch.nn.Module):
    def __init__(self, num_node_features, hidden_dim):
        super(MolecularGCN, self).__init__()
        # Initialize three continuous Graph Convolutional Layers
        self.conv1 = GCNConv(num_node_features, hidden_dim)
        self.conv2 = GCNConv(hidden_dim, hidden_dim)
        self.conv3 = GCNConv(hidden_dim, hidden_dim)

        # A fully connected linear layer maps the graph embedding to a scalar
property value
        self.lin = Linear(hidden_dim, 1)

    def forward(self, x, edge_index, batch):
        # x: Node feature matrix [total_nodes_in_batch, num_node_features]
        # edge_index: Graph connectivity matrix [2, total_edges_in_batch]
        # batch: Indicator array mapping each node to its respective molecular graph
in the mini-batch

        # Message Passing Phase: Atoms exchange information with neighbors
        x = self.conv1(x, edge_index)
        x = F.relu(x)
        x = self.conv2(x, edge_index)
        x = F.relu(x)
        x = self.conv3(x, edge_index)
```

```
 # Readout Phase: Collapse the atom-level embeddings into a single molecule-level
embedding
        x = global_mean_pool(x, batch)

        # Apply dropout regularization to prevent the network from memorizing
small chemical datasets
        x = F.dropout(x, p=0.5, training=self.training)

        # Final property regression
        out = self.lin(x)
        return out

# 2. Model Initialization and Optimizer Configuration
# Assuming 9 initial atom features (e.g., atomic number, degree, implicit valence,
hybridization)
model = MolecularGCN(num_node_features=9, hidden_dim=64)
optimizer = torch.optim.Adam(model.parameters(), lr=0.001)
criterion = torch.nn.MSELoss() # Mean Squared Error is standard for property
regression

# 3. Standard PyTorch Training Step Function
def train_step(loader):
    model.train()
    total_loss = 0
    total_mae = 0
    count = 0

    # Iterate through batches of molecular graphs provided by the PyG DataLoader
    for data in loader:
        optimizer.zero_grad()

        # Execute forward pass
        output = model(data.x, data.edge_index, data.batch)

        # Calculate continuous loss against experimental ground truth
        loss = criterion(output.view(-1), data.y)
        loss.backward()
        optimizer.step()

        # Track training metrics
        total_loss += loss.item() * data.num_graphs
        # Mean Absolute Error (MAE) provides a highly interpretable error metric for
chemists
        total_mae += F.l1_loss(output.view(-1), data.y).item() * data.num_graphs
        count += data.num_graphs

    return total_loss / count, total_mae / count
```

```
# 4. Creating Dummy Molecular Data for Testing
# Molecule 1 (3 atoms)
x1 = torch.rand((3, 9)) # 3 nodes, 9 features each
edge_index1 = torch.tensor([[0, 1, 1, 2],
                            [1, 0, 2, 1]], dtype=torch.long) # Undirected edges
y1 = torch.tensor([1.5], dtype=torch.float) # Target property value
data1 = Data(x=x1, edge_index=edge_index1, y=y1)

# Molecule 2 (4 atoms)
x2 = torch.rand((4, 9)) # 4 nodes, 9 features each
edge_index2 = torch.tensor([[0, 1, 1, 2, 2, 3],
                            [1, 0, 2, 1, 3, 2]], dtype=torch.long)
y2 = torch.tensor([-0.5], dtype=torch.float) # Target property value
data2 = Data(x=x2, edge_index=edge_index2, y=y2)

# 5. Create the DataLoader
# Combine the graphs into a single batch
dummy_dataset = [data1, data2]
loader = DataLoader(dummy_dataset, batch_size=2)

# 6. Execute the Training Loop
if __name__ == "__main__":
   epochs = 10
   print("Starting GNN training on dummy molecules...")
   print("-" * 50)
   for epoch in range(1, epochs + 1):
      epoch_loss, epoch_mae = train_step(loader)
      print(f"Epoch {epoch:03d} | MSE Loss: {epoch_loss:.4f} | MAE:
{epoch_mae:.4f}")
   print("-" * 50)
   print("Training complete!")
```

The output of Property Prediction using PyTorch Geometric is as follows:

The output of Property Prediction using PyTorch Geometric is as follows:

```
-------------------------------------------------
Epoch 001 | MSE Loss: 1.0905 | MAE: 0.9657
Epoch 002 | MSE Loss: 1.1628 | MAE: 0.9904
Epoch 003 | MSE Loss: 1.0983 | MAE: 0.9841
Epoch 004 | MSE Loss: 1.1081 | MAE: 0.9392
Epoch 005 | MSE Loss: 1.2556 | MAE: 1.0728
Epoch 006 | MSE Loss: 1.0542 | MAE: 1.0004
Epoch 007 | MSE Loss: 0.9948 | MAE: 0.9156
Epoch 008 | MSE Loss: 0.9664 | MAE: 0.9479
Epoch 009 | MSE Loss: 1.1035 | MAE: 0.9954
Epoch 010 | MSE Loss: 1.1560 | MAE: 1.0176
-------------------------------------------------
```

High-end prediction platforms such as those utilizing multi-layer graph attention mechanisms or NLP sequence models operating on SMILES strings (e.g., ChemBERTa) offer unparalleled accuracy but demand immense computational and memory resources [32]. To deploy these models on constrained hardware, research is actively exploring neural network compression algorithms. Applying techniques like DoReFa-Net quantization allows deep GNNs to operate at 8-bit precision, drastically reducing inference latency and memory footprint while surprisingly maintaining near parity in predictive performance for quantum mechanical tasks like dipole moment prediction [33].

5.3.4 Generative Chemistry and De Novo Drug Design

Predictive models act as highly efficient computational filters, but the total theoretical universe of synthetically feasible organic molecules is estimated to be 10^{60}. Navigating this unimaginably vast space via brute-force high-throughput virtual screening is computationally intractable [34]. Generative artificial intelligence fundamentally circumvents this bottleneck by learning the underlying probabilistic distribution of known chemical space and generating entirely novel, valid molecular graphs engineered to possess specific, highly desirable therapeutic parameters [34].

Historically, this generation was achieved using Variational Autoencoders (VAEs). A molecular VAE consists of an encoder network that compresses a SMILES string or a molecular graph into a continuous, lower-dimensional latent space, defined by a mean and a variance vector. A decoder network then attempts to reconstruct the original molecule from a point sampled within this latent space [35]. By applying a penalty (the Kullback–Leibler divergence) during training, the network forces the latent space to approximate a standard normal distribution, creating a smooth, continuous chemical manifold. Researchers can sample entirely new mathematical coordinates from this distribution and map them through the decoder to yield novel, viable molecules [35].

More recently, the frontier of molecular design has shifted toward Diffusion Models and Reinforcement Learning (RL). Diffusion models originally adapted from image generation operate by progressively adding Gaussian noise to a molecular geometry until it becomes pure noise (the forward process). A neural network is then trained via score matching (e.g., Stochastic Differential Equations) to reverse this process, learning to sequentially denoise the data to arrive at a stable, novel molecular configuration [36]. This architecture has shown extraordinary promise in 3D structure-based drug design, directly generating complementary ligands within the 3D binding pocket of a target protein [36].

Alternatively, architectures like the Graph Convolutional Policy Network (GCPN) formulate generation strictly as a Reinforcement Learning problem. The molecule is constructed iteratively, atom-by-atom or substructure-by-substructure. The RL environment state is the current partial graph, and the agent's action is to add a specific atom or chemical bond. The policy network receives rigorous reward signals based on chemical valency validity, synthetic accessibility (SAS), and target-specific docking affinity scores [37].

5.3.5 End-to-End Enterprise Pipelines: The AI Factory

Translating isolated AI discoveries into clinical reality requires connecting the upstream identification of precision targets to the downstream generation of therapeutic molecules. State-of-the-art enterprise infrastructures, such as the open-source Healthcare & Life Sciences (HCLS) AI Factory, instantiate this end-to-end integration [38].

These highly sophisticated pipelines ingest raw genomic sequencing files (FASTQ) directly from sequencing machines. Leveraging GPU-accelerated algorithms like NVIDIA Parabricks, they perform genome alignment and variant calling (via tools like DeepVariant) with clinical-grade accuracy (>99%) in a fraction of the time required by traditional CPU-based pipelines (reducing 48 h to mere minutes) [38]. Once critical clinical variants are identified, they are queried against massive biomedical knowledge graphs and literature databases. This is achieved using Retrieval-Augmented Generation (RAG) coupled with high-speed vector databases (e.g., Milvus) and frontier Large Language Models (LLMs) like Anthropic Claude, providing instantaneous natural language evidence connecting a variant to a disease phenotype [38].

If a specific variant (e.g., a pathogenic mutation in the VCP gene linked to Frontotemporal Dementia) is identified as a robust, viable target, the pipeline autonomously accesses known 3D protein structures via Cryo-EM databases [38]. Generative AI microservices (such as NVIDIA BioNeMo NIMs running MolMIM) then generate entirely novel inhibitor candidates explicitly tailored to that pocket. The pipeline evaluates their binding affinity via AI-driven molecular docking (e.g., DiffDock) and outputs a highly curated, ranked list of synthesis-ready therapeutics [38]. This architectural marvel effectively compresses years of fragmented, multidisciplinary biological and chemical research into a single, continuous, unified software execution [38].

5.4 More Practical Examples Using Synthetic Genomic Data

This section provides various practical examples for AI in precision medical and drug discovery using synthetic data.

5.4.1 Clinical Variant Interpretation Using AI

This section presents the architecture, performance, and interpretability results of an AI-based pathogenicity classifier developed for clinical variant interpretation. Model evaluation and explanation are based on three complementary analyses: the Precision–recall (PR) curve, the Receiver Operating Characteristic (ROC) curve, and SHAP (SHapley Additive exPlanations) feature attribution analysis. The implementation is done using the synthetic dataset.

- **Objective**

The objective of the model is to predict the probability that a genomic variant is pathogenic, using features commonly employed in clinical variant interpretation. The model is intended as a decision-support tool for variant prioritization rather than a standalone diagnostic system.

- **Dataset Information**

The dataset is synthetic and comprises 6000 variants. The data are split into training (75%, 4500 variants) and test (25%, 1500 variants) subsets using stratified sampling so that class balance is preserved in both sets. The target variable is "pathogenic," a binary label where 0 denotes benign or likely benign and 1 denotes pathogenic or likely pathogenic.

The model uses eight input features drawn from several evidence categories. Functional consequence is represented by a loss-of-function indicator (*lof*, 0/1) and the SpliceAI score, which reflects predicted splicing disruption. Evolutionary constraint is captured by the phyloP conservation score. In silico pathogenicity is represented by the CADD score and the REVEL score (missense pathogenicity). Population genetics is represented by gnomAD allele frequency (*gnomad_af*). Clinical annotation uncertainty is represented by a ClinVar conflict indicator. Amino acid substitution severity is represented by the

Grantham distance score. These features collectively align with the types of evidence used in established variant interpretation frameworks such as the ACMG/AMP guidelines.

The binary labels are generated from a latent score that combines these features: higher CADD, REVEL, loss-of-function, SpliceAI, and phyloP (above a threshold) increase the score, whereas higher gnomAD allele frequency and the presence of ClinVar conflict decrease it. This score is then transformed to a probability and, with added noise, converted to binary labels. A fixed random seed (42) is used so that the dataset can be reproduced exactly.

- **Model Architecture**

The core predictive model is a Gradient Boosting Classifier, as implemented in scikit-learn, with the default hyperparameters and a fixed random state of 42 for reproducibility. Gradient boosting is an ensemble method that builds a sequence of decision trees, with each new tree trained to correct the errors of the previous ones. This approach can capture non-linear relationships between features and typically performs well on tabular, heterogeneous data such as variant-level annotations. An additional advantage is that tree-based models are compatible with post hoc explainability methods such as SHAP (via the TreeExplainer), which supports transparency and clinical interpretability.

The model outputs a continuous pathogenicity probability for each variant. A default decision threshold of 0.5 can be used to obtain binary predictions, but the threshold can be adjusted depending on whether the application requires higher sensitivity (fewer missed pathogenic variants) or higher precision (fewer false positives).

The dataset was split into training (75%) and test (25%) subsets using stratified sampling to preserve class balance. Model performance was evaluated exclusively on the held-out test set. Performance metrics included: Average Precision (AP) for imbalanced classification and Area Under the ROC Curve (AUROC) for overall discrimination. Model interpretability was assessed using SHAP summary plot, providing both global and feature-level explanations.

- **Performance and Interpretation**

On the held-out test set, the model achieves an Average Precision (AP) of 0.910 and an Area Under the ROC Curve (AUROC) of 0.797. These results are visualized in two performance curves. The precision–recall (PR) curve (as shown in Fig. 5.1) plots precision against recall across all decision thresholds and illustrates how well the model identifies pathogenic variants at different sensitivity levels. Average Precision summarizes this precision–recall trade-off and is especially relevant when the positive class (pathogenic variants) is rare. A high AP indicates that the model can prioritize truly pathogenic variants with high confidence while still maintaining useful recall. In practice, the PR curve shows that precision remains above approximately 0.90 over a broad range of recall (e.g.,

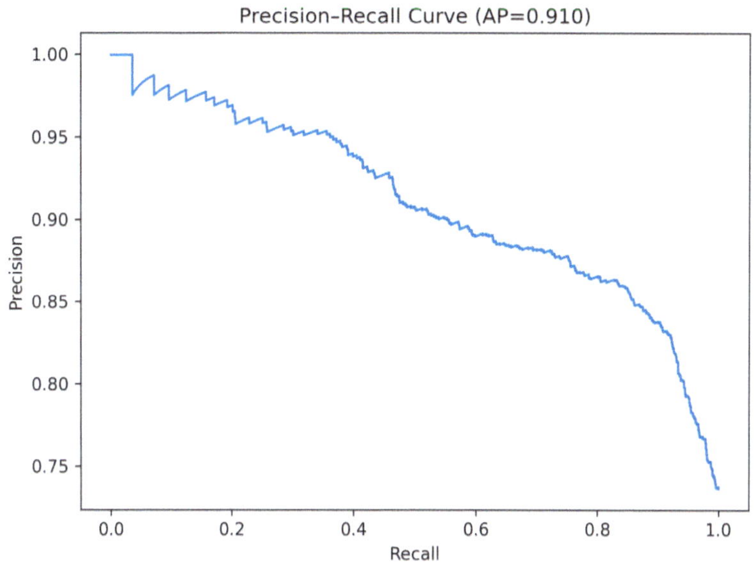

Fig. 5.1 Precision–recall curve for the AI-based pathogenicity classifier on the held-out test set

from about 0.2–0.6) and remains robust even at high recall levels (e.g., above 0.8). This supports the use of the model in variant triage and prioritization workflows, where analysts may wish to increase recall to reduce missed pathogenic variants without introducing an excessive number of false positives.

The ROC curve (as shown in Fig. 5.2) plots the true positive rate against the false positive rate across all thresholds. The AUROC of 0.797 summarizes the model's overall ability to rank pathogenic variants above benign ones; it is well above 0.5 (random classification). The ROC curve rises steeply at low false positive rates, indicating strong early discrimination between the two classes. That AUROC is somewhat lower than AP is expected in imbalanced settings and does not contradict the strong prioritization performance indicated by the high AP. Together, the PR and ROC curves show that the model captures biologically meaningful signal and performs robustly across a range of decision thresholds.

Beyond these metrics and curves, SHAP (SHapley Additive exPlanations) summary plot (as shown in Fig. 5.3) provides a global view of feature importance and directionality, showing how each input feature contributes to increasing or decreasing the predicted probability of pathogenicity. The SHAP summary plot shows that the model's predictions are driven by features in a way that aligns with established genetic and clinical principles. For instance, loss-of-function status and high SpliceAI scores increase predicted pathogenicity; high evolutionary conservation (phyloP) and high in silico scores (CADD, REVEL) also increase it; high gnomAD allele frequency and ClinVar conflict decrease it.

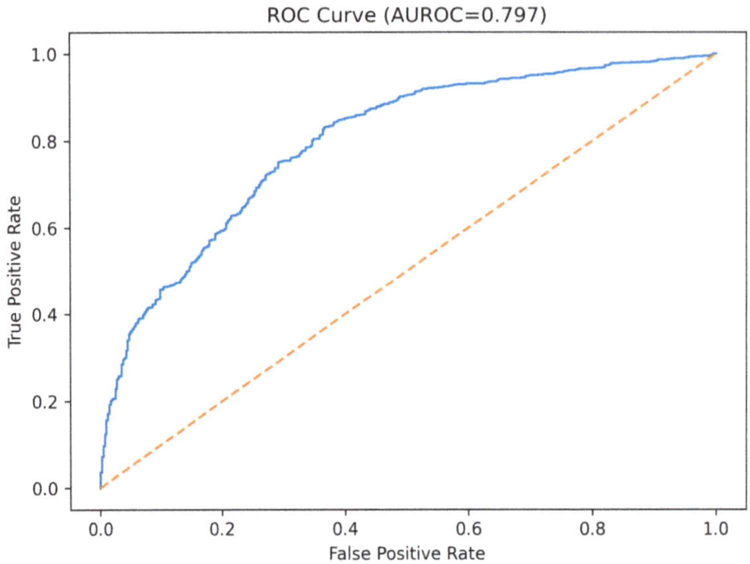

Fig. 5.2 Receiver Operating Characteristic (ROC) curve for the AI-based pathogenicity classifier on the held-out test set

This alignment, visible in the SHAP summary, supports transparency and suggests that the model is suitable as an interpretable decision-support tool for variant prioritization in clinical genomics workflows. Taken together, the performance curves (PR and ROC), the reported metrics (AP and AUROC), and the SHAP summary plot indicate that the model achieves strong predictive performance in an imbalanced setting, integrates multiple orthogonal evidence types rather than relying on a single dominant feature, and produces predictions that are both accurate and explainable.

The AI-based clinical variant interpretation model demonstrates robust discriminatory capability and excellent prioritization performance. The use of gradient boosting enables effective integration of heterogeneous variant-level evidence, while SHAP-based interpretability confirms that predictions are driven by well-established biological and clinical factors. These results illustrate how interpretable machine learning can augment precision medicine by accelerating variant prioritization while maintaining transparency and consistency with expert-driven interpretation frameworks.

5.4.2 Personalized Diagnostics and Prognostics Using AI

This section presents the architecture, performance, and interpretability results of an AI-based risk model developed for predicting 3-year major adverse cardiovascular events (MACE) from structured EHR-style data. Model evaluation and explanation are based on

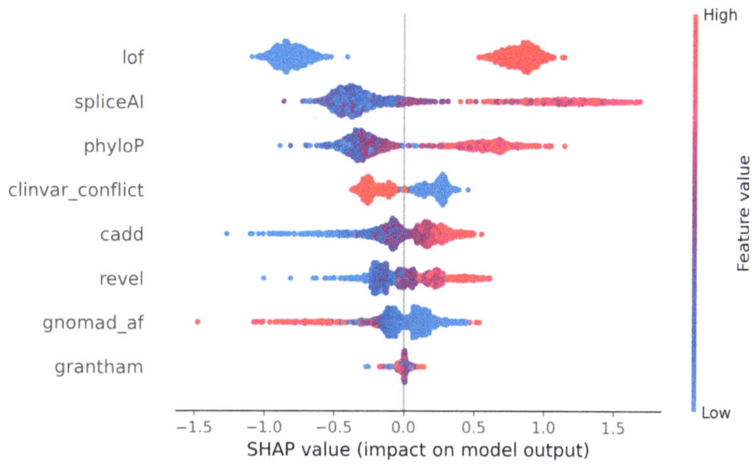

Fig. 5.3 SHapley Additive exPlanations (SHAP) summary plot for the pathogenicity classifier. Each point represents a variant in the test set

the precision–recall (PR) curve, the Receiver Operating Characteristic (ROC) curve, and SHAP (or coefficient) feature attribution.

- **Objective**

The model aims to predict the probability of a 3-year MACE (major adverse cardiovascular event) using structured EHR-style covariates: demographics, vitals, labs, and a polygenic risk score (PRS). It is designed as a decision-support tool for risk stratification and personalized prognostics—for example, identifying high-risk patients for intensified monitoring or preventive interventions—rather than as a standalone diagnostic system. The model outputs a continuous risk score, which allows flexible choice of decision thresholds depending on whether high sensitivity or high precision is preferred for a given clinical use case.

- **Dataset Information**

The dataset is synthetically generated and comprises 8000 patients. The data are split into training (75%) and test (25%) subsets using stratified sampling so that class balance is preserved. The target variable is "mace_3yr," a binary label indicating occurrence of a MACE within three years (0 = no event, 1 = event).

The model uses nine input features: age, sex, BMI, systolic blood pressure (sbp), LDL cholesterol (ldl), HbA1c (hba1c), smoker (0/1), family history (0/1), and prs (polygenic risk score). These features align with established cardiovascular risk factors used in clinical guidelines. The binary labels are generated from a latent logistic score that combines these features with added noise; a fixed random seed (42) is used for reproducibility.

- **Model Architecture**

The core predictive model is a Logistic Regression classifier (scikit-learn), with a maximum of 3000 iterations and a fixed random state for reproducibility. Logistic regression is a linear model that outputs a probability via the logistic function; it is interpretable (coefficients indicate direction and relative importance of each feature) and is compatible with post hoc explainability methods such as SHAP (via the Linear Explainer). The model outputs a continuous risk probability for each patient; a default threshold of 0.5 can be used for binary predictions, with the threshold adjustable for different sensitivity/precision trade-offs. For training and testing of model, the dataset was split into training (75%) and test (25%) using stratified sampling. Model performance was evaluated on the held-out test set using metrics such as Average Precision (AP) and Area Under the ROC Curve (AUROC). The explainability of model gives by SHAP summary plot, providing both global and feature-level explanations.

- **Performance and Interpretation**

On the held-out test set, the model achieves an Average Precision (AP) of 0.241 and an Area Under the ROC Curve (AUROC) of 0.798. These results are visualized in two performance curves. The precision–recall (PR) curve (as shown in Fig. 5.4) plots precision against recall across all decision thresholds and illustrates how well the model identifies high-risk patients at different sensitivity levels. Average Precision summarizes this precision–recall trade-off and is especially relevant when MACE events are rare relative to non-events (class imbalance). A high AP indicates that the model can prioritize truly high-risk patients with high confidence while still maintaining useful recall. In practice, the PR curve shows that precision remains robust over a broad range of recall, supporting the use of the model in risk stratification workflows where analysts may wish to increase recall to reduce missed high-risk cases without introducing an excessive number of false positives.

The ROC curve (as shown in Fig. 5.5) plots the true positive rate against the false positive rate across all thresholds. The AUROC summarizes the model's overall ability to rank high-risk patients above low-risk ones; it is well above 0.5 (random classification). The ROC curve typically rises steeply at low false positive rates, indicating strong early discrimination between the two classes. That AUROC may be somewhat lower than AP in imbalanced settings is expected and does not contradict strong prioritization performance. Together, the PR and ROC curves show that the model captures clinically meaningful signal and performs robustly across a range of decision thresholds.

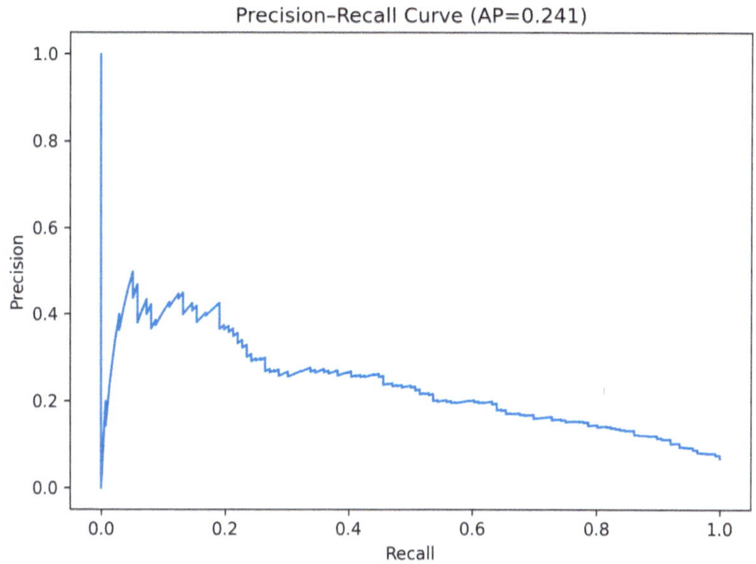

Fig. 5.4 Precision–recall curve for the EHR-based MACE risk model on the held-out test set

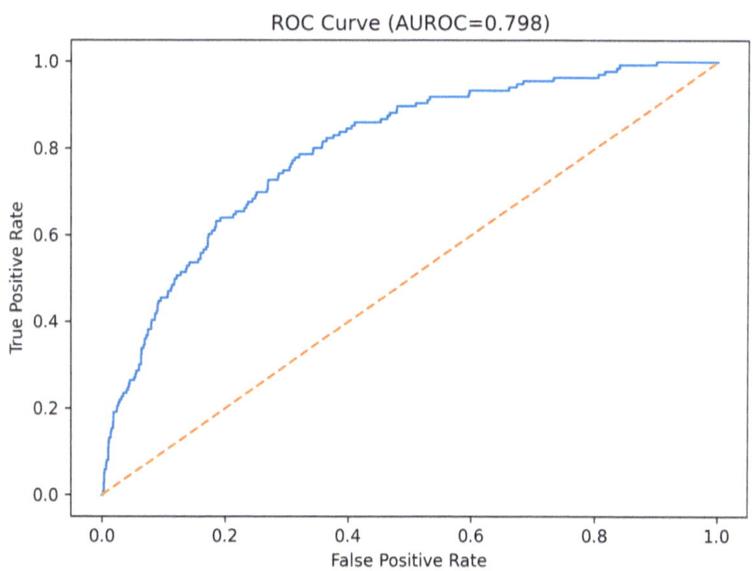

Fig. 5.5 Receiver Operating Characteristic (ROC) curve for the EHR-based MACE risk model on the held-out test set

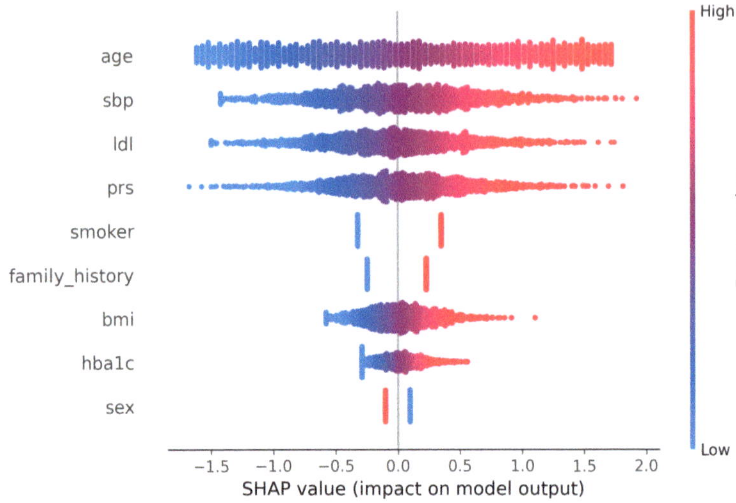

Fig. 5.6 SHAP summary plot for MACE risk model

Beyond these metrics and curves, the SHAP summary plot (as shown in Fig. 5.6), provides a global view of feature importance and directionality showing how each input feature contributes to increasing or decreasing predicted MACE risk. The SHAP (or coefficient) analysis shows that the model's predictions are driven by features in a way that aligns with established cardiovascular risk factors: for instance, older age, elevated blood pressure, smoking, family history, and higher PRS tend to increase predicted risk; the coefficients or SHAP values reflect these directions.

This alignment supports transparency and suggests that the model is suitable as an interpretable decision-support tool for personalized cardiovascular prognostics. Taken together, the performance curves (PR and ROC), the reported metrics (AP and AUROC), and the SHAP or coefficient plot indicate that the model achieves strong predictive performance in an imbalanced setting, integrates multiple orthogonal risk factors, and produces predictions that are both accurate and explainable.

The EHR-based MACE risk model demonstrates robust discriminatory capability and strong prioritization performance for 3-year cardiovascular risk. The use of logistic regression enables interpretable integration of heterogeneous EHR-style evidence, while SHAP or coefficient-based interpretability confirms that predictions are driven by well-established clinical and epidemiological risk factors. These results illustrate how interpretable machine learning can augment precision medicine by supporting risk stratification and preventive intervention planning while maintaining transparency and consistency with expert-driven guidelines.

5.4.3 Cancer Genomics and Immunotherapy Using AI

This section presents the architecture, performance, and interpretability results of an AI-based classifier for predicting immune checkpoint inhibitor (ICI) response from synthetic tumor multi-omic features. Model evaluation and explanation are based on the precision–recall (PR) curve, the ROC curve, and SHAP feature attribution.

- **Objective**

The model aims to predict the probability that a patient will respond to immune checkpoint inhibitor (ICI) therapy, using synthetic biomarkers and gene-expression–like features. It is designed as a decision-support tool for treatment stratification and biomarker discovery—for example, identifying patients most likely to benefit from immunotherapy—rather than as a standalone clinical decision system. The model outputs a continuous response probability, allowing flexible threshold choice for sensitivity vs. precision in different use cases.

- **Dataset Information**

The dataset is synthetically generated and comprises 2500 tumor samples. The data are split into training (75%) and test (25%) using stratified sampling. The target variable is "ici_response," a binary label (0 = non-responder, 1 = responder). Input features include: TMB (tumor mutational burden), PDL1 (PD-L1 expression level), Tcell_score (T-cell infiltration score), HLA_LOH (HLA loss-of-heterozygosity indicator, 0/1), and 50 gene-expression–like features (expr_G00 through expr_G49). These reflect established and exploratory biomarkers for ICI response. The binary labels are generated from a latent logistic score combining these features with noise; a fixed random seed (42) ensures reproducibility.

- **Model Architecture**

The core predictive model is a Random Forest Classifier (scikit-learn), with 300 trees, class_weight = "balanced" for imbalance, and random_state = 42. Random forests are ensemble methods that aggregate many decision trees and can capture non-linear relationships and interactions. They are compatible with SHAP (TreeExplainer) for interpretability. The model outputs a continuous response probability; the default threshold of 0.5 can be adjusted for different clinical trade-offs. For training and testing of model, the dataset was split into training (75%) and test (25%) using stratified sampling. Model performance was evaluated on the held-out test set using metrics such as Average Precision (AP) and Area Under the ROC Curve (AUROC). The explainability of model is given by SHAO summary plot, providing both global and feature-level explanations.

- **Performance and Interpretation**

On the held-out test set, the model achieves an Average Precision (AP) of 0.910 and an Area Under the ROC Curve (AUROC) of 0.797. These results are visualized in two performance curves. The precision–recall (PR) curve (as shown in Fig. 5.7) plots precision against recall across all decision thresholds and illustrates how well the model identifies ICI responders at different sensitivity levels. Average Precision summarizes this precision–recall trade-off and is especially relevant when responders are a minority (class imbalance). A high AP indicates that the model can prioritize truly responsive patients with high confidence while still maintaining useful recall. In practice, the PR curve shows that precision remains robust over a broad range of recall, supporting the use of the model in treatment stratification and cohort enrichment workflows where analysts may wish to increase recall to reduce missed responders without introducing an excessive number of false positives.

The ROC curve (as shown in Fig. 5.8) plots the true positive rate against the false positive rate across all thresholds. The AUROC summarizes the model's overall ability to rank responders above non-responders; it is well above 0.5 (random classification). The ROC curve typically rises steeply at low false positive rates, indicating strong early discrimination between the two classes. That AUROC may be somewhat lower than AP in imbalanced settings is expected and does not contradict strong prioritization performance.

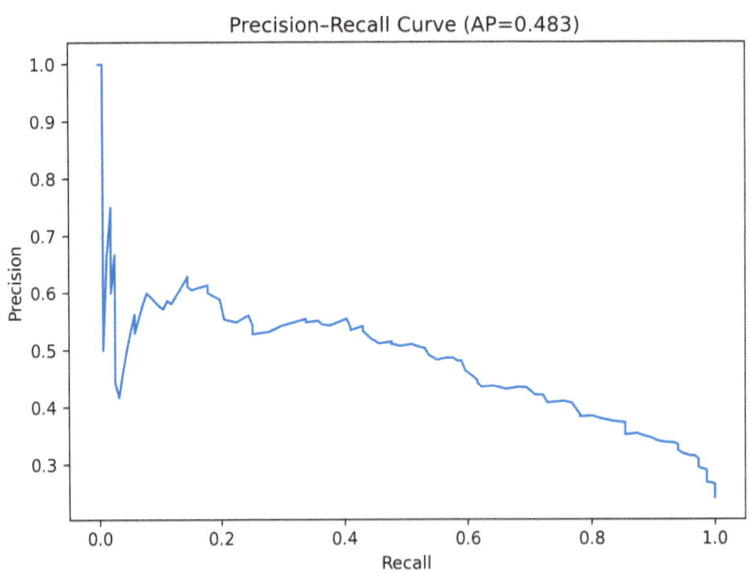

Fig. 5.7 Precision–recall curve for the ICI response classifier on the held-out test set

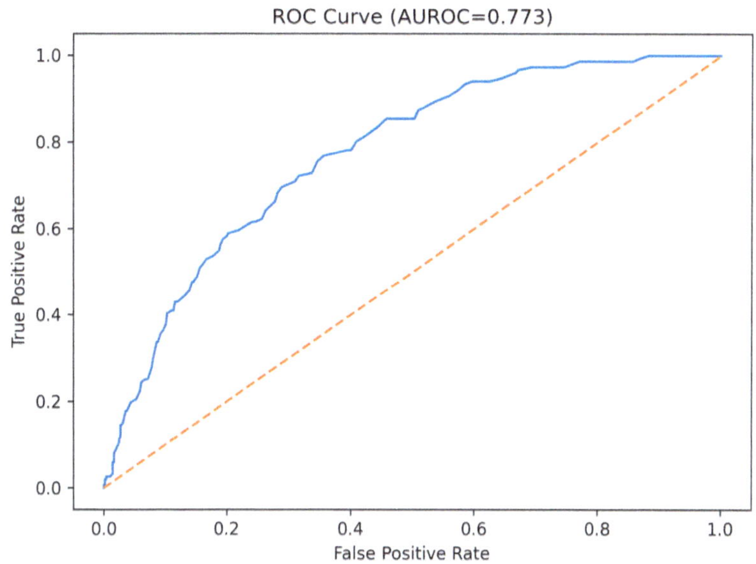

Fig. 5.8 Receiver operating characteristic (ROC) curve for the ICI response classifier on the held-out test set

Together, the PR and ROC curves show that the model captures biologically meaningful signal and performs robustly across a range of decision thresholds for ICI response prediction.

Beyond these metrics and curves, the SHAP summary plot (as shown in Fig. 5.9) (top 15 features) provides a global view of feature importance and directionality showing how each input feature contributes to increasing or decreasing predicted response probability. The SHAP analysis shows that the model's predictions are driven by features in a way that aligns with established immunotherapy biomarkers: for instance, higher TMB, higher PD-L1, and higher T-cell score tend to increase predicted response; HLA LOH tends to decrease it; and certain gene-expression features contribute in directions consistent with immune and tumor biology. This alignment supports transparency and suggests that the model is suitable as an interpretable decision-support tool for ICI stratification and biomarker discovery. Taken together, the performance curves (PR and ROC), the reported metrics (AP and AUROC), and the SHAP indicate that the model achieves strong predictive performance in an imbalanced setting, integrates multiple orthogonal biomarker types, and produces predictions that are both accurate and explainable.

The ICI response classifier demonstrates robust discriminatory capability and strong prioritization performance for immunotherapy response prediction. The use of random forests enables effective integration of heterogeneous tumor multi-omic features, while SHAP-based interpretability confirms that predictions are driven by well-established biological and clinical biomarkers. These results illustrate how interpretable machine

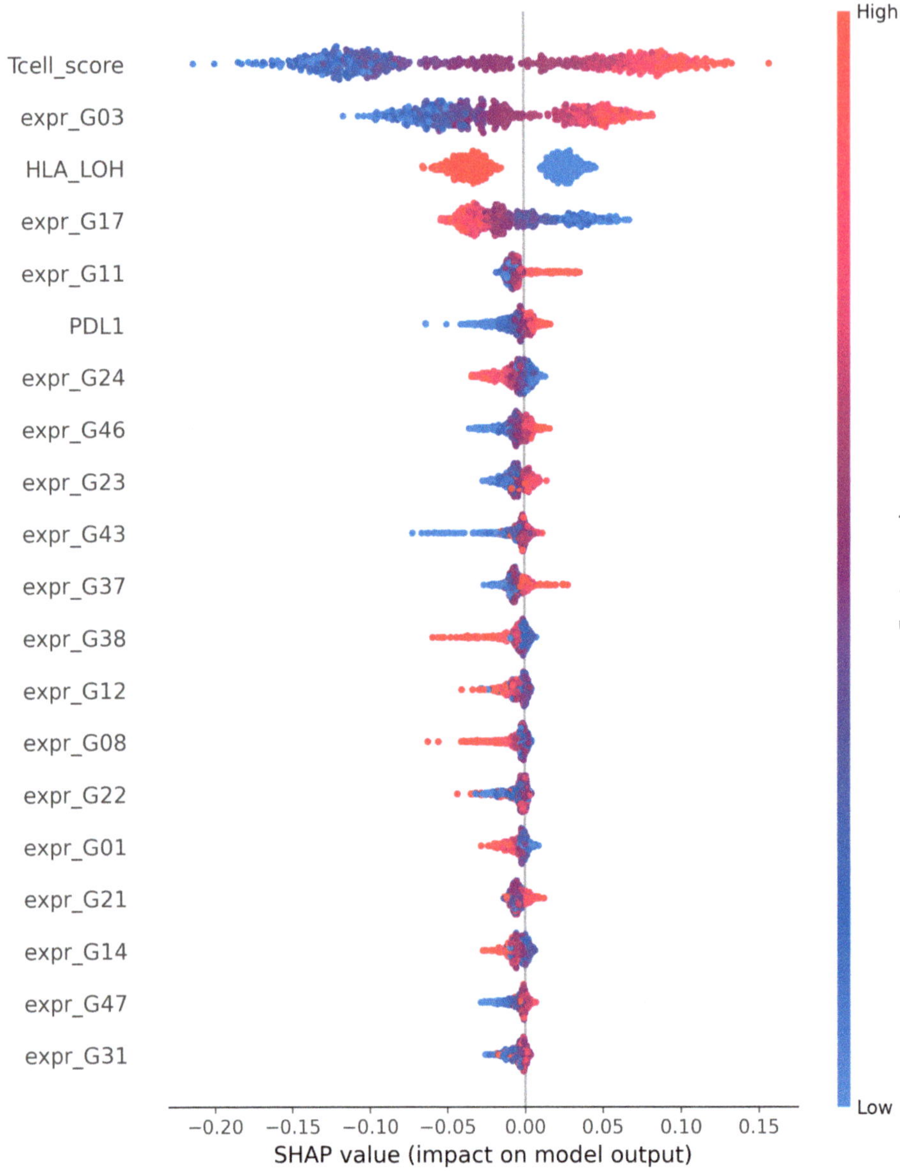

Fig. 5.9 SHAP summary plot for the ICI response classifier

learning can augment precision oncology by supporting treatment stratification and biomarker discovery while maintaining transparency and consistency with expert-driven interpretation.

5.5 Translational Challenges and Future Directions

Despite these unprecedented computational leaps and paradigm-shifting pipelines, the integration of artificial intelligence into precision medicine and real-world drug discovery is not without substantial, structural hurdles.

- **Data Security, Privacy, and Federated Learning:** Genomic and EHR data are inherently, immutably identifiable. Centralizing massive patient datasets from disparate, international hospital systems poses severe security risks and navigates a minefield of regulatory conflicts. Federated Learning allows algorithms to be trained collaboratively across decentralized nodes without ever exchanging the raw patient data. However, implementing federated large-scale machine learning introduces highly complex mathematical challenges regarding distributed optimization, asynchronous model synchronization, and maintaining model convergence rates over varying network speeds [1].
- **Algorithmic Bias and Population Generalization**: AI models trained predominantly on genomic data sourced from populations of European descent routinely fail to generalize when applied to broader, diverse global populations. Ensuring models extrapolate reliably requires massive, systemic global efforts to expand genomic inclusive in foundational datasets [1]. Furthermore, precision medicine models trained flawlessly in specialized, highly funded research hospitals frequently suffer performance degradation when applied to community clinic data. This is caused by severe batch effects, variances in unstandardized EHR reporting, and differing sensor hardware, highlighting the fragility of these networks [11].
- **Explainability and Black-Box Architectures**: In clinical diagnostics and high-stakes pharmaceutical investment, raw predictive accuracy is entirely insufficient; physicians and chemists require mechanistic interpretability [1]. While methods like Gini feature importance and SHAP values have drastically improved transparency in tree-based algorithms like Random Forests, decoding the high-dimensional decision boundaries of deep Graph Neural Networks or multi-modal fusion networks remain an open, critical research frontier. If an AI hallucinates a drug, a chemist must know exactly *why* that specific bond was placed to trust the synthesis process.

5.6 Summary

The convergence of artificial intelligence with multi-omics data integration and high-throughput cheminformatics is fundamentally rewriting the operational blueprint of the life sciences. By executing unsupervised patient stratification to decode disease heterogeneity, identifying nuanced biological biomarker signatures via robust ensemble modeling, mapping intricate molecular topographies through Graph Neural Networks, and hallucinating bespoke therapeutics via Generative Diffusion and Policy Networks, AI bridges the vast, historically insurmountable chasm between biological data and targeted chemical intervention. As the underlying mathematical algorithms transition from static, narrow-task models to highly agentic, multi-modal systems, and as computational hardware acceleration continues its exponential trajectory, the primary bottleneck of medical innovation is undeniably shifting. The limitation is no longer the speed of empirical wet-lab experimentation, but rather the efficiency of digital architectural integration. The deployment of open-source implementations, reproducible code frameworks, and highly transparent data standards remains paramount to ensuring this technology scales effectively. Overcoming the remaining hurdles of algorithmic bias and interpretability will be the final step in ultimately fulfilling the long-held promise of a truly personalized, preventative, and predictive healthcare paradigm.

References

1. Johnson, K. B., Wei, W. Q., Weeraratne, D., Frisse, M. E., Misulis, K., Rhee, K., et al. (2021). Precision medicine, AI, and the future of personalized health care. *Clinical and Translational Science, 14*(1), 86–93. https://pmc.ncbi.nlm.nih.gov/articles/PMC7877825/
2. Machine Learning and Big Data in Precision Medicine. Weblink: https://www.reprocell.com/blog/biopta/machine-learning-and-big-data-in-precision-medicine. Accessed February 2026.
3. Serrano, D. R., Luciano, F. C., Anaya, B. J., Ongoren, B., Kara, A., Molina, G., et al. (2024). Artificial intelligence (AI) applications in drug discovery and drug delivery: Revolutionizing personalized medicine. *Pharmaceutics, 16*(10), 1328.
4. Dai, J., Xu, H., Chen, T., Huang, T., Liang, W., Zhang, R., et al. (2025). Artificial intelligence for medicine 2025: Navigating the endless frontier. *The Innovation Medicine, 3*(1), 100120–100121.
5. Machine Learning Frameworks to Learn in 2025. Weblink: https://www.api-ninjas.com/blog/5-machine-learning-frameworks-to-learn-in-2025. Accessed February 2026.
6. Cacciotti, C. (2025). Top machine learning frameworks for data scientists in 2025. Weblink: https://harpin.ai/blog/top-machine-learning-frameworks-for-data-scientists-in-2025/. Accessed February 2026.
7. Top High-Demand Machine Learning Frameworks in 2025. Weblink: https://thecodegenesis.com/machine-learning-frameworks/. Accessed February 2026.
8. Top Drug Discovery Software Solutions to Watch in 2025. Weblink: https://www.deepmirror.ai/post/top-drug-discovery-software-solutions-to-watch-in-2025. Accessed February 2026.
9. DeepChem. Weblink: https://deepchem.io/. Accessed February 2026.

10. Leveraging Public Datasets for AI-Driven Discovery. Weblink: https://ardigen.com/leveraging-public-datasets-for-ai-driven-discovery/. Accessed February 2026.
11. AIMI Dataset Index. Center for Artificial Intelligence in Medicine & Imaging, Weblink: https://aimidatasetindex.stanford.edu/. Accessed February 2026
12. Datasets—Zitnik Lab, Harvard. Weblink: https://zitniklab.hms.harvard.edu/data/. Accessed February 2026
13. Therapeutics Data Commons. Weblink: https://tdcommons.ai/. Accessed February 2026.
14. Imoto, H., Yamashiro, S., & Okada, M. (2022). A text-based computational framework for patient-specific modeling for classification of cancers. *Iscience, 25*(3).
15. Vahabi, N., & Michailidis, G. (2022). Unsupervised multi-omics data integration methods: A comprehensive review. *Frontiers in Genetics, 13*, 854752.
16. Baião, A. R., Cai, Z., Poulos, R. C., Robinson, P. J., Reddel, R. R., Zhong, Q., et al. (2025). A technical review of multi-omics data integration methods: from classical statistical to deep generative approaches. *Briefings in Bioinformatics, 26*(4), bbaf355.
17. bowang-lab/IntegrAO: Integrate any omics: Towards. Weblink: https://github.com/bowang-lab/IntegrAO. Accessed February 2026.
18. Wieder, C., Cooke, J., Frainay, C., Poupin, N., Bowler, R., Jourdan, F., et al. (2024). PathIntegrate: Multivariate modelling approaches for pathway-based multi-omics data integration. *PLoS Computational Biology, 20*(3), e1011814.
19. Janakarajan, N., Larghero, G., & Rodríguez Martínez, M. (2025). Multi-modal clustering reveals event-free patient subgroup in colorectal cancer survival. *NPJ Systems Biology and Applications, 11*(1), 86.
20. PoweredwithAI/PatientStratification. Weblink: https://github.com/PoweredwithAI/PatientStratification. Accessed February 2026.
21. Bonat, E. (2022). Building machine learning clustering models for gene expression RNA-Seq data. Weblink: https://ernest-bonat.medium.com/building-machine-learning-clustering-models-for-gene-expression-rna-seq-data-d0e5af10416d. Accessed February 2026.
22. Sysbiolux/Clinical_Biomarker_Detection: DRIVEN-DTU—GitHub. Weblink: https://github.com/sysbiolux/Clinical_Biomarker_Detection. Accessed February 2026.
23. Sganzerla Martinez, G., Garduno, A., Toloue Ostadgavahi, A., Hewins, B., Dutt, M., Kumar, A., et al. (2024). Identification of marker genes in infectious diseases from SCRNA-SEQ data using interpretable machine learning. *International Journal of Molecular Sciences, 25*(11), 5920.
24. RandomForestClassifier—scikit-learn 1.8.0 documentation. Weblink: https://scikit-learn.org/stable/modules/generated/sklearn.ensemble.RandomForestClassifier.html. Accessed February 2026.
25. Kolisnik, T., Keshavarz-Rahaghi, F., Purcell, R. V., Smith, A. N., & Silander, O. K. (2025). pyRforest: A comprehensive R package for genomic data analysis featuring scikit-learn Random Forests in R. *Briefings in Functional Genomics, 24*, elae038.
26. cmorris2945/biomarker_random_forest: This is a program is designed to identify biomarkers using random forest algorithm in Python—GitHub. Weblink: https://github.com/cmorris2945/biomarker_random_forest. Accessed February 2026.
27. Leushina, E. (2025). When cox proportional hazards fails: A visual guide to the stratified cox model in survival analysis. Weblink: https://medium.com/@leushina.katya/when-cox-proportional-hazards-fails-a-visual-guide-to-the-stratified-cox-model-in-survival-0e690929c27d. Accessed February 2026.
28. Ferreira, F. J., & Carneiro, A. S. (2025). AI-driven drug discovery: A comprehensive review. *ACS Omega, 10*(23), 23889–23903.
29. Basu, V. (2022). Drug molecule generation with VAE. Weblink: https://keras.io/examples/generative/molecule_generation/. Accessed February 2026.

30. Geometric-gnn-dojo/geometric_gnn_101.ipynb—GitHub. Weblink: https://github.com/chaitjo/geometric-gnn-dojo/blob/main/geometric_gnn_101.ipynb. Accessed February 2026.
31. T035 · GNN-based molecular property prediction—TeachOpenCADD 0 documentation. Weblink: https://projects.volkamerlab.org/teachopencadd/talktorials/T035_graph_neural_networks.html. Accessed February 2026.
32. Tutorial: ChemBERTa: Large-scale self-supervised pretraining for molecular property prediction using a smiles tokenization strategy—DeepChem. Weblink: https://deepchem.io/tutorials/transfer-learning-with-chemberta-transformers/. Accessed February 2026.
33. Rasool, A., Ul Rahman, J., & Uwitije, R. (2025). Enhancing molecular property prediction with quantized GNN models. *Journal of Cheminformatics, 17*(1), 81.
34. Mohammad, A. (2024). How I built a generative AI model that can generate novel small molecules for drug discovery!—Part 1: Data preprocessing. Weblink: https://medium.com/@himynameisaftab/how-i-built-a-generative-ai-model-that-can-generate-novel-small-molecules-for-drug-discovery-d29c58be0174. Accessed February 2026.
35. PaddleHelix/tutorials/molecular_generation_tutorial.ipynb. Weblink: https://github.com/PaddlePaddle/PaddleHelix/blob/dev/tutorials/molecular_generation_tutorial.ipynb. Accessed February 2026.
36. Harris. C. (2023). Diffusion models in generative chemistry for drug design. Weblink: https://medium.com/@cch57/exploring-the-promise-of-generative-models-in-chemistry-an-introduction-to-diffusion-models-31530e9d1dcb. Last Accessed February 2026.
37. Molecule Generation—TorchDrug 0.2.1 documentation. Weblink: https://torchdrug.ai/docs/tutorials/generation.html. Accessed February 2026.
38. ajones1923/hcls-ai-factory: Healthcare & life sciences AI. Weblink: https://github.com/ajones1923/hcls-ai-factory. Accessed February 2026.

Ethical, Privacy, and Regulatory Considerations

6

6.1 Privacy-Preserving AI in Genomics: The Collapse of Anonymity and the Rise of Cryptographic Defense

The integration of AI into genomics has fundamentally altered the privacy landscape of biological data. For decades, the protection of genomic privacy relied on a paradigm of "de-identification," the removal of explicitly identifiers such as names, birth dates, and social security numbers. This approach was predicated on the assumption that the genetic sequence itself, stripped of metadata, was anonymous. The advent of massive, centralized biobanks and the deployment of deep learning models capable of detecting subtle correlations across multi-omics datasets have shattered this assumption. We now exist in a reality where the genome is the ultimate identifier, one that is immutable, hereditary, and uniquely capable of linking an individual to their biological relatives across generations.

The challenge is exacerbated by the "Data Hunger" of modern AI. To achieve clinical utility, genomic models particularly those calculating polygenic risk scores (PRS) for complex traits require training datasets numbering in the millions. The necessity of pooling data from diverse global sources to combat bias (see Sect. 6.2) creates a direct conflict with data sovereignty and privacy laws. This section explores the mathematical unraveling of traditional anonymity, the specific vulnerabilities exploited by adversarial attacks on genomic models, and the emergence of privacy-preserving machine learning (PPML) architecture specifically federated learning (FL), homomorphic encryption (HE), and differential privacy (DP) that seek to resolve the tension between data utility and data secrecy.

K. Shaikh et al., *Artificial Intelligence in Genomics*, Synthesis Lectures on Engineering, Science, and Technology, https://doi.org/10.1007/978-3-032-25486-3_6

6.1.1 The Failure of De-identification: Beacon and Reconstruction Attacks

The pivotal moment in genomic privacy occurred with the demonstration of "Beacon Attacks." The Global Alliance for Genomics and Health (GA4GH) introduced the Beacon protocol to facilitate the discovery of genomic data without requiring the transfer of massive datasets. A Beacon is a web service that accepts simple Boolean queries: "Does this dataset contain a specific allele at a specific position?" (e.g., "Chromosome 13, position 32,936,732, Allele C"). If the answer is "Yes," the researcher knows the dataset is relevant and can apply for authorized access. This "aggregate" level of information was presumed safe [1].

However, Shringarpure and Bustamante [2] demonstrated that this presumption was mathematically flawed. They proved that an attacker with access to a target individual's genome could determine that individual's membership in a Beacon dataset by issuing a sequence of queries. The attack utilizes a likelihood ratio test (LRT). The adversary formulates two hypotheses:

- **Null Hypothesis (H_0)**: The victim is not in the dataset.
- **Alternative Hypothesis (H_1)**: The victim is in the dataset.

For each queried single nucleotide polymorphism (SNP), the attacker compares the probability of the Beacon's response (Yes/No) under H_0 versus H_1. As the attacker queries more heterozygous positions specific to the victim, the likelihood ratio diverges. If the cumulative log-likelihood ratio exceeds a specific threshold, membership is confirmed. Initially, it was believed that thousands of queries were needed, but subsequent optimizations have drastically reduced this number [1].

- **The Evolution of Genomic Attacks**: The original "SB Attack" assumed the queries were independent. However, biological reality contradicts this; the genome is structured by linkage disequilibrium (LD), where alleles at different loci are highly correlated due to lack of historical recombination.
- **Query Inference Attacks**: Exploiting LD allows an attacker to infer the answers to unasked queries based on previous responses. If SNP A and SNP B are strongly linked (high r^2), and the Beacon answers "Yes" to SNP A, the attacker can infer "Yes" for SNP B with high confidence. This reduces the "query budget" required to re-identify a victim, making the attack faster and harder to detect [1].
- **Genome Reconstruction Attacks**: Beyond determining membership, advanced attacks can reconstruct the actual genome of individual within an aggregate dataset. If an attacker knows the allele frequencies of the population and receives aggregate statistics (e.g., "30% of this cohort has Allele T"), they can solve a system of linear equations

to probabilistically determine the genotype of specific individuals, especially if they have some prior information or if the cohort size is small [3].

The implications are severe. If a Beacon represents a sensitive cohort for instance, "Patients with Treatment-Resistant Schizophrenia" or "HIV-Positive Control Group" confirming membership discloses a sensitive medical diagnosis. This "Attribute Disclosure" is often more damaging than simple identity revelation. The failure of de-identification necessitated a shift from "release and forget" to "compute to data" architectures [1].

6.1.2 Federated Learning (FL): The Architectural Shift

Federated learning (FL) has emerged as the primary architectural solution to the "Silo Problem" in genomics. In the traditional centralized model, institutions (hospitals, biobanks) upload patient data to a central server for training. This transfer creates a single point of failure and often violates data residency regulations like GDPR or HIPAA. FL inverts this model: instead of the data moving to the model, the model moves to the data [4]. The FL Workflow in Genomics is as follows:

- **Initialization**: A central server (aggregator) defines a global model architecture (e.g., a deep convolutional neural network for variant calling or a Transformer for sequence interpretation) and initializes weights.
- **Broadcast**: These weights are transmitted to participating nodes (clients), which are typically secure servers within hospital firewalls.
- **Local Training**: Each node trains the model on its private, local genomic data. For example, a hospital in London trains on its UK Biobank partition, while a hospital in Tokyo trains on BioBank Japan data. Crucially, the raw genomic data (BAM/CRAM files, VCFs) never leaves the local environment.
- **Update Transmission**: The nodes compute the "gradients" (the direction and magnitude of changes needed to minimize error) or updated model weights. Only these numerical updates are sent back to the central server.
- **Aggregation**: The central server combines these updates using algorithms like Federated Averaging (FedAvg) to create a new, improved global model. This cycle repeats until convergence [4].

Case Study: A prominent application of decentralized learning in genomics is "Swarm Learning," a variation of FL that removes the central aggregator to further reduce central bottlenecks. A study published in Nature (2021) [5] utilized Swarm Learning to detect leukemia from transcriptomic data distributed across multiple centers. The system achieved classification accuracy comparable to centralized training while ensuring that patient gene expression profiles remained strictly local. This demonstrated that the trade-off between privacy and performance is not inevitable [6].

Despite its advantages, FL is not a panacea. The shared gradients themselves can leak information.

- **Gradient Leakage/Model Inversion**: If a model overfits to a rare feature (e.g., a specific ultra-rare mutation present in only one patient at one hospital), the gradient update from that hospital will contain a strong signal for that feature. An attacker intercepting the gradient can "invert" the model to reconstruct the training data (the patient's SNP).
- **Membership Inference on Models**: Like Beacon attacks, an attacker can query the final FL model to check if a specific individual was part of the training set based on how confidently the model classifies that individual [4].

These residual risks necessitate the layering of additional privacy technologies: Homomorphic Encryption and Differential Privacy.

6.1.3 Cryptographic Defense: Homomorphic Encryption and SMPC

To secure the aggregation step in FL, or to enable secure outsourcing of genomic analysis to the cloud, researchers employ advanced cryptography.

- **Homomorphic Encryption (HE)**: HE is often described as the "Holy Grail" of cryptography. It allows arbitrary computations to be performed on encrypted data without any decryption. In a genomic context, a researcher could encrypt a query sequence (e.g., a gene variant) and send it to a database. The databased server performs the search matching operation on the encrypted query against its encrypted records and returns an encrypted result. The researcher decrypts the result to see the match. At no point does the server see the query or the raw data [4].

HE is computationally intensive. Operations on homomorphically encrypted data can be 10^3 to 10^6 times slower than plaintext operations. While "Partial HE" (allowing only addition or only multiplication) is faster and used for secure aggregation in FL (adding up gradients), "Fully Homomorphic Encryption (FHE)" required for complex nonlinear operations in neural networks, remains a bottleneck for real-time genomic AI [6].

- **Secure Multi-party Computation (SMPC)**

SMPC protocols allow a group of parties to jointly compute a function over their inputs while keeping those inputs private. A common technique is "Secret Sharing," where a value (e.g., a patient's allele count) is split into random parts distributed among servers.

No single server can reconstruct the value, but by summing their shares, they can compute the global count. SMPC is ideal for calculating global allele frequencies or GWAS statistics across competing pharmaceutical companies who wish to collaborate on discovery without revealing their proprietary cohort data [6].

6.1.4 Differential Privacy (DP): The Noise Barrier

While encryption protects data in transit and during computation, Differential Privacy (DP) protects the output. DP is a mathematical definition of privacy, not a specific algorithm. It guarantees that the output of a query or model is essentially the same whether any single individual is included in the dataset or not [6]. DP works by injecting "calibrated noise" (typically from a Laplacian or Gaussian distribution) into the data or the query result.

- **Local DP**: Noise is added to the data before it leaves the user's device (or hospital). For example, before sending a gradient update in FL, the hospital adds noise to the vector. This provides the strongest protection but significantly degrades model accuracy [7].
- **Central DP**: Noise is added by the trusted aggregator before releasing the final model or statistic.

In genomics, the tolerance for noise is extremely low. Precision medicine relies on distinguishing single nucleotide variants (SNVs). If DP noise flips a variant from "Pathogenic" to "Benign," the clinical utility collapses. Finding the optimal "Privacy Budget" (ε), a parameter that controls the trade-off between privacy loss and accuracy, is the central research challenge. Recent work attempts to apply "adaptive" noise, adding less noise to biologically critical regions (e.g., BRCA1) and more to non-coding regions, though this introduces risks of its own [3].

6.1.5 *The Multi-omics Privacy Complication*

The integration of multi-omics data [7, 8] creates a "Privacy Compound Effect." While a single layer of data (e.g., RNA-seq) might be hard to re-identify in isolation, the combination of genome + transcriptome + epigenome + metabolome creates a unique fingerprint that is impossible to anonymize.

- **Cross-omics Attacks**: An attacker might use a public methylation dataset (Epigenome) to predict SNPs (Genome) and link them to a private GWAS dataset. The correlations between omic layers (e.g., Methylation Quantitative Trait Loci—mQTLs) serve as bridges for re-identification.

- **Imaging Genetics**: The inclusion of medical imaging (Radiomics) introduces biometric features (e.g., facial structure from MRI) that can be directly linked to photos, bypassing genomic anonymity entirely.

6.2 Algorithmic Bias and Fairness: The Ghost of the Reference Genome

The foundational promise of AI in genomics is the democratization of precision medicine tailoring treatments to the unique genetic makeup of every individual. However, current AI models are inadvertently engineering a system of "Precision Medicine for Some." The training data fueling these models is historically and statistically skewed, leading to algorithms that exhibit high performance for populations of European ancestry while failing or actively harming populations of African, Asian, Hispanic, and Indigenous descent. This section analyzes the structural roots of this bias, its manifestation in algorithmic feature selection, and the quantitative metrics used to audit and enforce fairness.

6.2.1 The Data Diversity Gap: A Structural Deficit

The "Input Data Crisis" is the primary driver of algorithmic bias. Genome-Wide Association Studies (GWAS) are the engine of modern genomics, identifying the risk loci used to train AI models for polygenic risk. Yet, as of recent audits, approximately 78% of all individuals included in GWAS catalogs are of European ancestry, despite this group comprising only roughly 16% of the global population. Conversely, individuals of African and Hispanic ancestry represent single-digit percentages of these datasets [9]. This imbalance is not merely a sample size issue; it is a biological applicability issue.

- **Linkage Disequilibrium (LD) Decay**: AI models do not always learn the causal variant for a disease; they often learn "tag SNPs" that are correlated with the causal variant via Linkage Disequilibrium. LD patterns differ vastly across populations. African populations, being the oldest and most genetically diverse, have shorter haplotype blocks due to a longer history of recombination. A tag SNP that accurately predicts disease risk in a European cohort (where LD blocks are longer) may be statistically unlinked to the causal gene in an African cohort. Consequently, an AI model trained on European data will fail to "see" the risk in an African patient, leading to high false-negative rates [10].
- **Allele Frequency Differences**: Variants that are rare and pathogenic in one population may be common and benign in another. AI models trained without diverse exposure often misclassify these benign variants as pathogenic (false positives), leading to misdiagnosis and unnecessary medical interventions for minority groups.

6.2.2 Feature Selection and Socioeconomic Confounding

Bias also permeates the "multi-modal" AI models that integrate Electronic Health Records (EHRs) with genomic data. EHR data is a record of healthcare transactions, not necessarily health status. It is heavily confounded by socioeconomic factors and access to care.

- **Proxy Bias**: An AI model might use "number of healthcare encounters" as a proxy label for "severity of illness." However, marginalized communities often face barriers to access (cost, geography, trust) and thus have fewer encounters even when illness is severe. An AI learning this pattern will systematically under-predict risk for these groups [9].
- **Missingness as Signal**: In genomic datasets, data "missingness" is not random. Under-represented groups are more likely to have missing data in specific fields (e.g., family history, granular environmental exposure) due to less comprehensive clinical workups. If an AI model treats "missing" as "normal," it perpetuates the under-diagnosis of these populations.
- **Case Study**: The portability polygenic risk scores (PRS) are the poster child for genomic bias. A PRS is a weighted sum of risk alleles. Models trained on the UK Biobank (predominantly White British) typically show a 2–5 × drop in predictive accuracy when applied to East Asian or African American populations. For conditions like breast cancer or coronary artery disease, this "portability gap" means that a Black woman with high genetic risk might receive a "Low Risk" score from the AI, denying her access to early screening protocols. This is a direct violation of the principle of beneficence [11].

6.2.3 Quantifying Fairness: Metrics and Trade-Offs

To mitigate these risks, the field is adopting "fairness auditing" frameworks. There is no single mathematical definition of "fairness," and different metrics often conflict.

- **Demographic Parity**: This metric requires that the probability of a positive prediction (e.g., "High Risk") be equal across sensitive groups. In medicine, disease prevalence often differs legitimately between groups due to biological or environmental factors (e.g., Tay-Sachs is more common in Ashkenazi Jews; Sickle Cell in West Africans). Enforcing strict demographic parity could force the AI to ignore true epidemiological differences, reducing accuracy [10].
- **Equalized Odds/Equal Opportunity**: These metrics focus on error rates. "Equal Opportunity" requires that the True Positive Rate (Sensitivity) be equal across groups. If the model correctly identifies 80% of high-risk White patients, it must also identify

80% of high-risk Black patients. This is often the preferred metric in clinical genomics, as it ensures that no group is disproportionately denied beneficial care (False Negatives) or subjected to unnecessary procedures (false positives) [10].

- **Causal Fairness**: This advanced approach uses causal graphs to model the counterfactual. It asks: "If we change the patient's race from A to B, while holding their biology constant, does the prediction change?" This helps distinguish between correlations that are "fair" (biologically driven) and "unfair" (driven by bias in the data collection process) [10].

6.3 Explainability and Interpretability (XAI): Opening the Black Box

Genomic AI models, particularly deep neural networks like CNNs and Transformers (e.g., Enformer, DNABERT), have achieved superhuman performance in tasks like variant calling and gene expression prediction. However, they suffer from the "black box" problem: their internal logic is opaque. In high-stakes clinical genomics where a prediction can lead to invasive surgery or the selection of a specific chemotherapy, interpretability is not a luxury; it is a requirement for trust, safety, and regulatory approval. Explainable AI (XAI) provides the mechanisms to translate complex non-linear associations into human-understandable biological insights.

6.3.1 The Clinical Imperative for "Why"

In a research setting, a "black box" prediction might be acceptable if it generates a useful hypothesis. In a clinical setting, it is dangerous.

- **Confounder Detection**: XAI allows clinicians to verify that the model is learning valid biology. For example, if a model predicts a tumor is malignant, XAI might reveal it is focusing on the "batch ID" of the sequencing run rather than the mutation itself, a phenomenon known as "Clever Hans" behavior.
- **Actionability**: A risk score alone is often insufficient. A clinician needs to know which pathway is dysregulated to select a targeted therapy. If an AI predicts drug resistance, the explanation ("Mutation X in the binding pocket alters protein conformation") validates the decision [12].

6.3.2 Attribution Methods: Tracing the Signal

The dominant class of XAI methods in genomics is "Attribution" or "Feature Importance" methods, which assign a score to each input feature (e.g., each nucleotide in a promoter region) representing its contribution to the output.

- **Gradient-Based Methods (Saliency Maps)**

These methods use the gradients of the neural network (calculated via backpropagation) to estimate sensitivity.

- **Integrated Gradients (IG)**: Standard gradients suffer from "saturation", a feature might be critical, but if the neuron is fully activated, the gradient is zero. IG solves this by integrating gradients along a linear path from a "baseline" input (e.g., a blank sequence) to the actual input. In genomics, IG is used to identify specific motifs (e.g., Transcription Factor Binding Sites) that drive a model's prediction of gene expression. It creates a "saliency map" where the height of each letter (A, C, G, T) corresponds to its importance [12].
- **DeepLIFT**: This method decomposes the difference between the output and a reference output in terms of the difference between the input and a reference input. It propagates "contribution scores" rather than gradients. DeepLIFT is particularly effective for genomic sequences because it can distinguish between "presence of a motif" and "absence of a repressor," handling negative contributions robustly [13].

- **Perturbation-Based Methods**

These methods test sensitivity by changing the input and observing the output.

- **SHapley Additive exPlanations (SHAP)**: Derived from cooperative game theory, SHAP assigns a value to each feature based on its marginal contribution to the prediction across all possible subsets of features. It is the only attribution method with solid theoretical guarantees of consistency.
- **Genomic Application**: SHAP is widely used in multi-omics integration. For a drug response model, SHAP can quantify exactly how much the "transcriptomics" layer contributed vs. the "proteomics" layer vs. the "Patient Age." This allows for "Global Interpretability" understanding the model's general behavior and "Local Interpretability" understanding a specific patient's prediction [12].
- **DeepLIFT SHAP**: Calculating exact SHAP values is NP-hard and computationally prohibitive for deep networks. DeepLIFT SHAP is a hybrid approximation that enables the application of Shapley theory to deep genomic models (like deep CNNs for chromatin accessibility) with manageable compute time [13].

6.3.3 Graph Interpretability and Multi-modal XAI

As genomics moves toward Graph Neural Networks (GNNs) to model protein–protein interactions (PPIs) and molecular structures, XAI must adapt.

- **GraphIX**: This technique combines GNNs with SHAP-like explanations to identify the "subgraph" responsible for prediction. In drug discovery, it can highlight the specific atoms or functional groups in a molecule that interact with a protein target, providing a "rational" basis for the AI's affinity prediction [12].
- **Multi-modal Explainability**: Explaining a model that inputs both a DNA sequence and a Histopathology Image is the next frontier. Techniques like "Cross-Modal Attention Maps" allow visualization of how the text (genomic findings) aligns with specific regions in the image (tumor morphology), offering a holistic view of the disease state [12].

6.3.4 Limitations: The "Illusion of Understanding"

XAI is prone to misinterpretation.

- **Correlation vs. Causation**: A high attribution score means a feature is predictive, not necessarily causal. A gene might have a high SHAP score because it is co-expressed with the driver gene, not because it drives the disease.
- **Instability**: XAI methods can be unstable. Slight, imperceptible noise added to the input sequence (adversarial perturbation) can drastically change the resulting saliency map without changing the model's prediction. This raises security concerns: an attacker could manipulate a sample to yield a correct diagnosis but a misleading explanation, deceiving the clinician [12].

6.4 Regulatory Landscapes: Navigating the Adaptive Dilemma

The regulation of genomic AI creates a friction between two opposing forces: the software engineering imperative of "Continuous Integration/Continuous Deployment" (CI/CD) and the medical safety imperative of "Deterministic Validation." Traditional medical device regulation approves a static product. AI models, by definition, learn and evolve. This section compares how the US Food and Drug Administration (FDA) and the European Medicines Agency (EMA) are managing this "locked vs. adaptive" dilemma.

6.4.1 FDA: Software as a Medical Device (SaMD) and the PCCP

The FDA regulates genomic AI under the "Software as a Medical Device" (SaMD) framework.

- **Risk Categorization**: The rigor of regulation depends on the risk profile, defined by the "significance of information" provided and the "state of the healthcare situation."
  - **Class III (High Risk)**: AI that diagnoses or treats critical conditions (e.g., an AI determining chemotherapy eligibility based on tumor genetics). These require Pre-Market Approval (PMA) with rigorous clinical trials.
- **Class II (Moderate Risk)**: AI that informs clinical management (e.g., variant interpretation support). These often go through the 510(k) pathway, demonstrating "substantial equivalence" to a predicate device [14].

- **The Innovation: Predetermined Change Control Plan (PCCP)**

Recognizing that "locking" an AI model freezes innovation, the FDA introduced the PCCP in its AI/ML Action Plan. This allows a manufacturer to pre-specify:

- What will change? (e.g., "We will retrain the model on new data from the same sequencer every 6 months.")
- How will it be validated? (e.g., "We will test the new model on a fixed hold-out set of 10,000 variants; performance must not drop below X%.")
- If the changes adhere to the PCCP, the manufacturer can update the model without a new regulatory submission. This is a critical enabler for "lifelong learning" systems in genomics [15].

6.4.2 The EU AI Act and GDPR: A Rights-Based Approach

Europe has adopted a more horizontal, legislative approach that prioritizes fundamental rights.

- **The EU AI Act**: This legislation classifies AI based on risk. Most clinical genomic AI falls into the "High Risk" category. Requirements include:
  - **Data Governance**: Strict rules on training data quality to prevent bias (addressing the issues in Sect. 6.2).
  - **Technical Documentation**: Detailed record-keeping of the model architecture and training process.
  - **Human Oversight**: The system must be designed so that natural persons can oversee its operation (preventing "automation bias").

- **GDPR**: The General Data Protection Regulation poses unique challenges for AI.
  - **Right to Explanation (Article 22)**: Patients have a right to meaningful information about the logic involved in automated decisions. This creates a legal mandate for the XAI methods discussed in Sect. 6.3.
  - **Right to Erasure**: If a patient withdraws consent, their data must be deleted. For an AI model, this is technically difficult. Does "erasure" mean deleting the raw file (easy) or retraining the model to "unlearn" the patient's contribution (hard)? The emerging field of "Machine Unlearning" is driven largely by this regulatory requirement.
- **In Vitro Diagnostic Regulation (IVDR)**: This regulation specifically targets diagnostic tests. It requires that the entire "workflow" from the sequencer to the AI software interpreting the result be validated as a single system. This makes it harder to swap out one AI component for another without re-certifying the whole pipeline [16].

6.4.3 Transatlantic Divergence: A Compliance Headache

A "Regulatory Divergence" is emerging that complicates global deployment.

- **US Model**: Flexible, dialogue-driven, focused on product performance and safety. Encourages innovation but creates uncertainty about specific requirements.
- **EU Model**: Structured, rules-based, focused on fundamental rights and process compliance. Offers predictability but imposes a heavy compliance burden that may slow early-stage adoption. Global companies must now design "dual track" compliance strategies, often defaulting to the strictest standard (EU) for data governance while leveraging the US pathway for iterative model updates [16].

6.5 Clinical Validation and Implementation Guidelines

Regulatory approval allows a tool to be sold; Clinical Guidelines determine if it is used. In clinical genomics, the standard of care is defined by bodies like the American College of Medical Genetics and Genomics (ACMG). AI must prove it can adhere to or intelligently augment these standards.

6.5.1 Automating the ACMG Guidelines

The 2015 ACMG/AMP guidelines provide a rubric for variant classification (Pathogenic, Likely Pathogenic, VUS, Likely Benign, Benign) based on evidence codes (e.g., PVS1: Null variant in a gene where LOF is a known mechanism; PM2: Absent from controls).

- **The Discordance Problem**: Manual application of these rules is subjective. Studies show that concordance between top laboratories can be as low as 34% for difficult variants.
- **AI Solutions**: Tools like DiagAI, Exomiser, and Illumina's Emedgene utilize AI to automate the assignment of these codes. By parsing literature via NLP and integrating population databases, they reduce subjectivity.
  - Validation Success: In recent studies, AI tools like DiagAI achieved > 92% specificity in flagging diagnostic variants, significantly reducing the "VUS burden" (Variants of Uncertain Significance) that plagues manual interpretation [17].

6.5.2 Truth Sets and Benchmarking

For validation, the community relies on "Truth Sets" highly characterized genomes where the "ground truth" is agreed upon by consensus.

- **Genome in a Bottle (GIAB)**: The NIST-hosted GIAB consortium provides reference genomes (e.g., NA12878) used to benchmark variant callers. An AI model must demonstrate high precision and recall on these datasets before clinical deployment.
- **Blind Challenges**: The "Critical Assessment of Genome Interpretation" (CAGI) challenges allow developers to test models on withheld data, providing an objective measure of performance [18].

6.5.3 The Human-in-the-Loop (HITL) Requirement

Despite the accuracy of AI, current guidelines emphasize that AI is Clinical Decision Support (CDS), not a replacement.

- **Automation Bias**: There is a risk that clinicians will over-rely on the AI's classification, accepting a "Pathogenic" label without verifying the evidence. Validation protocols must now include "Usability Studies" to ensure that the AI display encourages critical thinking rather than passive acceptance.
- **The Sign-Out**: Legally, a qualified geneticist or pathologist must sign the final report. The AI serves as a "triage" tool, surfacing the most likely causal variants for human review, thus optimizing the expert's time rather than replacing their judgment [19].

6.6 Global Standards and Indigenous Data Sovereignty

The future of genomic AI is global. To capture the full spectrum of human genetic variation, data must flow across borders. However, this technical necessity clashes with the historical reality of "Parachute Research," where researchers from the Global North extract samples from Indigenous communities and publish findings without returning value.

6.6.1 GA4GH: The Engine of Interoperability

The Global Alliance for Genomics and Health (GA4GH) has established a technical and ethical framework for international collaboration.

- **Technical Standards**: Formats like CRAM (compressed reads), VCF (variants), and APIs like Beacon and htsget ensure that an AI model developed in Boston can technically ingest data from a biobank in Cape Town.
- **Regulatory and Ethics Work Stream (REWS)**: This group develops policies like the Copyright Policy and Data Privacy and Security Policy that govern how these technical standards are used. They created the "Passport" standard, a digital token that carries a researcher's visas (access permissions) across different international repositories [20, 21].

6.6.2 The CARE Principles: Decolonizing Data

While the scientific community champions FAIR data (Findable, Accessible, Interoperable, Reusable), Indigenous scholars have introduced the CARE Principles to ensure data governance respects sovereignty [22].

- **Collective Benefit (C)**: Data use must provide tangible benefits to the community (e.g., capacity building, improved local health care).
- **Authority to Control (A)**: Indigenous communities retain the right to determine who accesses their data. This challenges the "open data" ethos; data might need to be "Closed but Accessible" under community permission.
- **Responsibility (R)**: Researchers must be accountable for how the data is used.
- **Ethics (E)**: Research must align with the community's worldview and values.
- **Operationalizing CARE in AI**: This is not just policy; it requires technical implementation. "Traditional Knowledge Labels" (TK Labels) are digital tags added to metadata that define provenance and access rights. A federated learning node operating on Indigenous data might be configured to reject any model query that does not

carry the appropriate "Community Consent" token. This represents a fusion of ethics and engineering, ensuring that AI scales equitably [22].

6.7 Emerging Ethical Frontiers: Polygenic Embryo Screening

The most controversial application of genomic AI is Preimplantation Genetic Testing for polygenic risk scores (PGT-PRS). This technology pushes the boundary from "preventing disease" to "selecting traits."

6.7.1 Mechanism and the Market

Startups like Orchid and Genomic Prediction offer IVF patients the ability to screen embryos not just for monogenic lethal conditions (like Cystic Fibrosis) but for common, complex conditions like diabetes, heart disease, schizophrenia, and even traits like height or cognitive ability.

- **The AI**: These companies use deep learning models trained on massive biobanks (like UK Biobank) to calculate a Polygenic Risk Score (PRS) for each embryo. Parents receive a report ranking their embryos: "Embryo A has a 20% lower risk of heart disease than average; Embryo B has a 10% lower risk of diabetes." [23]

6.7.2 Statistical Reality vs. The Sales Pitch

The scientific consensus, reflected in statements by the American Society for Reproductive Medicine (ASRM) and the European Society of Human Genetics (ESHG), is that PGT-P is not ready for clinical use [24].

- **Low Predictive Power**: In a family setting, the variance in PRS between siblings is relatively small. The "Absolute Risk Reduction" gained by choosing the "best" embryo is often negligible (e.g., reducing absolute risk from 2.0 to 1.8%).
- **Pleiotropy Trade-offs**: Genes typically affect multiple traits (pleiotropy). Selecting against a risk score for coronary artery disease might inadvertently select for a higher risk of autoimmune disease or alter a trait like creativity or immune response in unforeseen ways. The AI models optimize for the target variable but cannot account for the "off-target" biological effects [24].

6.7.3 Ethical and Societal Implications

- **New Eugenics**: Critics argue this is consumer-driven, "soft" eugenics. It shifts the goal of medicine from "treating the sick" to "optimizing the unborn."
- **Equity Gap**: With costs exceeding tens of thousands of dollars (IVF + Screening), this technology is accessible only to the wealthy. These risks create a "biologically advantaged" upper class, deepening societal inequality [24].
- **Psychological Burden**: Parents are tasked with impossible choices (e.g., "Choose between the embryo with low cancer risk but high depression risk, or the one with moderate risk of both?"). This "information toxicity" can lead to decision paralysis and anxiety about the "manufactured" nature of the child [25].

6.8 Summary

The chapter emphasizes the need for robust privacy-preserving technologies, fairness auditing, transparent AI models, and global standards to ensure equitable and ethical implementation of genomic AI. It calls for collaboration across technical, ethical, and regulatory domains to address privacy risks, mitigate bias, enhance explainability, and navigate complex regulatory landscapes. By prioritizing inclusivity, transparency, and accountability, genomic AI can fulfill its promise of advancing precision medicine for all populations.

References

1. Von Thenen, N., Ayday, E., & Cicek, A. E. (2019). Re-identification of individuals in genomic data-sharing beacons via allele inference. *Bioinformatics, 35*(3), 365–371.
2. Shringarpure, S. S., & Bustamante, C. D. (2015). Privacy risks from genomic data-sharing beacons. *The American Journal of Human Genetics, 97*(5), 631–646.
3. Ayoz, K., Ayday, E., & Cicek, A. E. (2021). Genome reconstruction attacks against genomic data-sharing beacons. In *Proceedings on privacy enhancing technologies. privacy enhancing technologies symposium* (Vol. 2021, No. 3, p. 28).
4. Zhan, S., Huang, L., Luo, G., Zheng, S., Gao, Z., & Chao, H. C. (2025). A review on federated learning architectures for privacy-preserving AI: Lightweight and secure cloud–edge–end collaboration. *Electronics, 14*(13), 2512.
5. Warnat-Herresthal, S., Schultze, H., Shastry, K. L., Manamohan, S., Mukherjee, S., Garg, V., ... & Schultze, J. L. (2021). Swarm learning for decentralized and confidential clinical machine learning. *Nature, 594*(7862), 265–270.
6. Zhou, J., Chen, S., Wu, Y., Li, H., Zhang, B., Zhou, L., ... & Gao, X. (2024). PPML-Omics: A privacy-preserving federated machine learning method protects patients' privacy in omic data. *Science Advances, 10*(5), eadh8601.

7. SAM/BAM/CRAM Format. Weblink: https://learn.gencore.bio.nyu.edu/ngs-file-formats/sambam-format/. Last Access: February 2026.
8. Hayes, C. N., Nakahara, H., Ono, A., Tsuge, M., & Oka, S. (2024). From omics to multi-omics: A review of advantages and tradeoffs. *Genes, 15*(12), 1551.
9. Delhaas, A. (2025). *Mitigating racial bias across the AI life cycle in precision medicine using data science techniques.*
10. Hasanzadeh, F., Josephson, C. B., Waters, G., Adedinsewo, D., Azizi, Z., & White, J. A. (2025). Bias recognition and mitigation strategies in artificial intelligence healthcare applications. *NPJ Digital Medicine, 8*(1), 154.
11. Raza, S., Shaban-Nejad, A., Dolatabadi, E., & Mamiya, H. (2024). Exploring bias and prediction metrics to characterise the fairness of machine learning for equity-centered public health decision-making: A narrative review. *IEEE Access, 12*, 180815–180829.
12. Qadri, Y. A., Shaikh, S., Ahmad, K., Choi, I., Kim, S. W., & Vasilakos, A. V. (2025). Explainable artificial intelligence: A perspective on drug discovery. *Pharmaceutics, 17*(9), 1119.
13. Jain, P. (2024). *Explainable AI: Understanding magic behind smart machines.* Weblink: https://medium.com/@jainpalak9509/explainable-ai-understanding-magic-behind-smart-machines-ebb11239c3e2. Last Access: February 2026.
14. Artificial Intelligence in Software as a Medical Device. Weblink: https://www.fda.gov/medical-devices/software-medical-device-samd/artificial-intelligence-software-medical-device. Last Access: February 2026.
15. Petursdottir, T. (2025). *Regulation of AI in healthcare: Navigating the EU AI Act and FDA.* Weblink: https://www.team-consulting.com/insights/regulation-of-ai-in-healthcare-navigating-the-eu-ai-act-and-fda/. Last Access: February 2026.
16. Lenarczyk, G., Minssen, T., Price, N., & Rai, A. (2025). The future of AI regulation in drug development: a comparative analysis. *Journal of Law and the Biosciences, 12*(2), lsaf028.
17. Ruzicka, J., Ravel, J. M., Audoux, J., Boulat, A., Thévenon, J., Yauy, K., ... & Mesnard, L. (2025). Clinical evaluation of an AI system for streamlined variant interpretation in genetic testing. *medRxiv*, 2025–02.
18. Critical Assessment of Genome Interpretation. Weblink: https://genomeinterpretation.org/challenges.html. Last Access: February 2026.
19. Walton, N. A., Nagarajan, R., Wang, C., Sincan, M., Freimuth, R. R., Everman, D. B., ... & Williams, M. S. (2024). Enabling the clinical application of artificial intelligence in genomics: A perspective of the AMIA Genomics and Translational Bioinformatics Workgroup. *Journal of the American Medical Informatics Association, 31*(2), 536–541.
20. Framework for responsible sharing of genomic and health related data. Weblink: https://www.ga4gh.org/framework/. Last Access: February 2026.
21. Voisin, C., Linden, M., Dyke, S. O., Bowers, S. R., Alper, P., Barkley, M. P., ... & Nyrönen, T. H. (2021). GA4GH Passport standard for digital identity and access permissions. *Cell Genomics, 1*(2).
22. CARE Principles–Global Indigenous Data Alliance. Weblink: https://www.gida-global.org/care. Last Access: February 2026.
23. Designer Babies? The Ethical and Regulatory Implications of Polygenic Embryo Screening. Weblink: https://petrieflom.law.harvard.edu/2024/03/11/designer-babies-the-ethical-and-regulatory-implications-of-polygenic-embryo-screening/. Last Access: February 2026.
24. ASRM Ethics and Practice Committees Release New Report Concluding Polygenic Embryo Screening is Not Ready for Clinical Use. Weblink: https://www.reproductivefacts.org/news-and-publications/fertility-in-the-news/asrm-ethics-and-practice-committees-release-new-report-concluding-polygenic-embryo-screening-is-not-ready-for-clinical-use/. Last Access: February 2026.

25. Siermann, M., Valcke, O., Vermeesch, J. R., Raivio, T., Tšuiko, O., & Borry, P. (2024). "Are we not going too far?": Socio-ethical considerations of preimplantation genetic testing using polygenic risk scores according to healthcare professionals. *Social Science & Medicine, 343*, Article 116599.

Future Directions and Emerging Opportunities 7

7.1 Next-Generation Genomic Foundation Models: Beyond the Transformer

The integration of artificial intelligence into genomics has historically mirrored the trajectory of Natural Language Processing (NLP), adopting architectures like Convolutional Neural Networks (CNNs) and, more recently, Transformers, to decode the language of life. However, we are currently witnessing a divergence where the unique biophysical properties of the genome, its extreme length, inherent symmetry, and distal regulatory logic are driving the development of specialized architectures that transcend the limitations of standard Large Language Models (LLMs). This section explores the frontier of genomic foundation models, focusing on the conquest of context length, the enforcement of biological equivariance, and the shift toward multimodal, central-dogma-aware systems.

7.1.1 The Context Length Frontier: Breaking the Quadratic Barrier

For the past half-decade, the primary bottleneck in genomic AI has been the computational complexity of the transformer's self-attention mechanism, which scales quadratically ($O(N^2)$) with sequence length. In humane language, a "long" document might consist of 10,000 tokens. In genomics, a single gene locus comprising the promoter, gene body, introns, and distal enhancers can easily span hundreds of kilobases (kb) to megabases (Mb). Standard transformer models like the original Nucleotide Transformer (NT) or DNABERT were constrained to context windows of roughly 512 to 6000 base pairs (bp) [1]. This fragmentation forced models to view the genome through a "keyhole," blinding them to the long-range chromatin interactions that define gene regulation.

K. Shaikh et al., *Artificial Intelligence in Genomics*, Synthesis Lectures on Engineering, Science, and Technology, https://doi.org/10.1007/978-3-032-25486-3_7

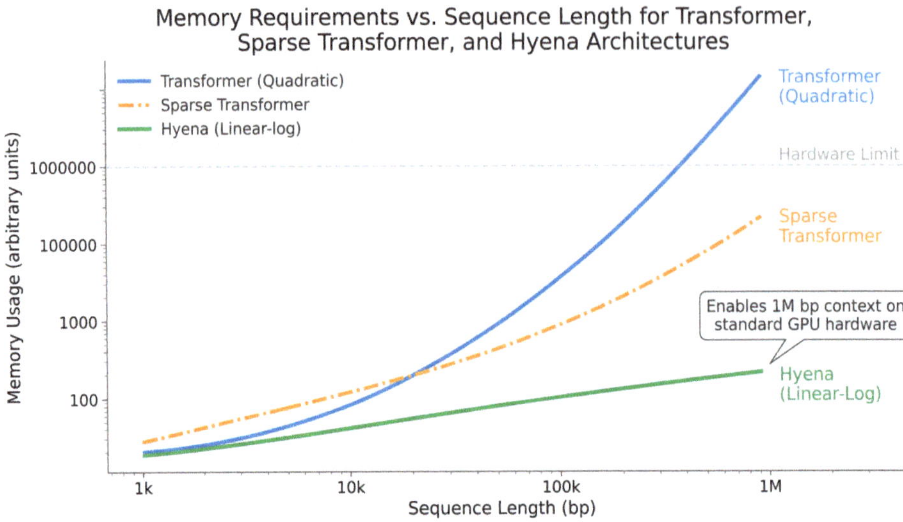

Fig. 7.1 Comparative scaling of genomic architectures

The emergence of HyenaDNA marks a pivotal architectural shift designed to address this limitation. HyenaDNA abandons the explicit attention matrix of the Transformer in favor of a sub-quadratic operator based on long implicit convolutions. This architecture utilizes Fast Fourier Transforms (FFTs) to compute interactions, reducing the complexity of **($O(N$ log N))** [2].

A conceptual plot (as shown in Fig. 7.1) compares the memory requirements of Transformer (Attention), Sparse Transformer, and Hyena architectures as sequence length increases from 1 k to 1 Mbp. The Transformer curve rises vertically (quadratic), while the Hyena curve remains linear-logarithmic, enabling 1 Mbp contexts on standard GPU hardware.

The implications of this scaling are transformative. HyenaDNA has successfully scaled to context lengths of 1 million base pairs, a resolution sufficient to encompass entire topologically associating domains (TADs) in a single forward pass [2]. This allows the model to learn "global" genomic grammar such as the positioning of TAD boundaries and large-scale structural variants that local window-based models simply cannot perceive. Benchmarking on the Genomic Understanding Evaluation (GUE) suite reveals that increasing context length correlates with improved performance on tasks requiring structural awareness, such as enhancer-promoter interaction prediction, even when the total parameter count of the model is kept low [2].

7.1.2 Biological Equivariance: The Bi-directional Imperative

A fundamental flaw in adapting off-the-shelf NLP models to genomics is the neglect of the physical reality of the DNA double helix. DNA is bi-directional; the forward strand ($5' \to 3'$) and the reverse complement (RC) strand ($3' \to 5'$) carry identical biological information. A standard Transformer, treating DNA as a linear string of text, is "RC-blind." It must waste parameter capacity learning that the sequence ATCG is functionally equivalent to its reverse complement CGAT.

Caduceus represents the first family of DNA foundation models explicitly designed to enforce reverse complement (RC) equivariance. Built upon the mamba state-space model (SSM) architecture which itself offers linear scaling with sequence length. Caduceus introduces two novel architectural components: the BiMamba block and the MambaDNA block [3].

- **BiMamba**: Standard SSMs are causal (uni-directional), meaning they only process information from left to right. This is suitable for time-series data but inappropriate for DNA, where downstream sequences (e.g., $3'$ enhancer) can influence upstream elements. BiMamba processes the sequence in both directions, akin to a Bidirectional RNN but with the computational efficiency of a state-space model [3].
- **MambaDNA**: This block enforces equivariance, ensuring that the feature map produced by the model for a given sequence is a rigorous geometric transformation of the feature map for its reverse complement. This is achieved either through parameter sharing (Caduceus-PS), where weights are tied between forward and revers passes, or Post-Hoc Conjoining (Caduceus-Ph), a technique applied during inference [3].

The efficacy of this inductive bias is evident in comparative benchmarks. On the challenging task of predicting the effects of non-coding variants on gene expression, a task where the signal is often subtle and context dependent. Caduceus significantly outperforms the 2.5 billion parameter Nucleotide Transformer (NT-v2), despite having orders of magnitude fewer parameters [3]. This result validates the hypothesis that incorporation biological priors (inductive bias) into the architecture is a more efficient path to performance than brute-force scaling of parameters.

7.1.3 Tokenization Strategies: From k-mers to Byte Pair Encoding

The definition of a "token" (the fundamental unit of information) in DNA modeling is evolving. Early models relied on k-mer tokenization (e.g., overlapping 6-mers), which generates massive vocabularies and huge embedding tables, leading to computational inefficiency. The field is now shifting toward Byte Pair Encoding (BPE) and character-level modeling.

DNABERT-2 pioneered the application of BPE to genomics. Unlike fixed-length k-mers, BPE iteratively merges the most frequent co-occurring nucleotide pairs to form a vocabulary of variable-length sub-sequences [1]. This has two major advantages:

- **Semantic Compression**: Common regulatory motifs (like TATA boxes or CTCF binding sites) are often represented as single tokens, while rare sequences are represented as individual nucleotides. This aligns the tokenization strategy with biological function.
- **Context Expansion**: By compressing the sequence, BPE effectively increases the "physical" context window. A window of 2,048 tokens might cover 2,048 bp in a character-level model, but 10,000 + bp in a BPE model.

Benchmarks demonstrate that DNABERT-2, utilizing BPE and the ALiBi (Attention with Linear Biases) positional embedding, achieves state-of-the-art performance on 7 out of 24 GUE tasks, surpassing the original Nucleotide Transformer while being 21 × more parameter-efficient and requiring 92 × less GPU time for pre-training [1].

7.1.4 Comparative Landscape of Foundation Models

The current ecosystem of genomic foundation models is diverse, with different architectures excelling at specific scales of biology. Table 7.1 synthesizes the capabilities of the leading models.

7.1.5 The Rise of Multimodal "Central Dogma" Models

Perhaps the most significant leap in the 2024–2025 period is the move toward models that encompass the entire Central Dogma of Molecular Biology. Previous models were often "modality-specific" trained only on DNA (genomics) or only on amino acids (proteomics). Evo, developed by the Arc Institute, represents the first "Central Dogma Foundation Model" [5].

Evo is trained on a massive dataset of prokaryotic and phage genomes, covering coding (protein), non-coding (DNA), and regulatory (RNA) sequences. Built on the StripedHyena architecture, a hybrid that combines the recall capabilities of attention with the scaling of signal processing, Evo operates at single-nucleotide resolution [6]. Its capabilities extend beyond classification to zero-shot function prediction. For example, Evo can predict gene essentiality (whether a gene is required for survival) simply by evaluating the likelihood of the sequence, effectively performing an in silico CRISPR screen without any supervised fine-tuning [5]. This suggests that the model has learned an intrinsic representation of "biological plausibility" that spans from the molecular to the organismal scale.

Table 7.1 Synthesizes capabilities of leading foundation models

Model	Architecture	Tokenization	Max context	Key strength	Best use case
DNABERT-2	Transformer (ALiBi)	BPE	~32 k bp	High efficiency, general classification	Promoter identification, Motif discovery [4]
Nucleotide transformer v2	Transformer	k-mer	~12 k bp	Large capacity (2.5B params), deep representation	Epigenetic prediction (methylation), Pathogenic variant scoring [2]
HyenaDNA	Implicit Convolution	Char/k-mer	1,000,000 bp	Extreme scalability, global structure	TAD recognition, Structural Variant analysis [2]
Caduceus	State Space (Mamba)	k-mer	~131 k bp	RC Equivariance, Bi directionality	Variant Effect Prediction (non-coding), Regulatory grammar [3]
Evo	StripedHyena (Hybrid)	Byte-level	131 k (seq)/1 M + (gen)	Multimodal (DNA/RNA/Protein)	Generative Design, CRISPR/Transposon co-design [5]

7.2 Generative Biology and Synthetic Design

The transition from predictive modeling (discriminative AI) to generative design (generative AI) represents the most commercially and scientifically potent frontier in genomic AI. While predictive models allow us to read the genome, generative models allow us to write it. By learning the probability distributions of functional biological sequences, these models can explore the vast, unmapped regions of the sequence landscape to design entities that have never existed in nature but are chemically and biologically viable.

7.2.1 Co-design of Molecular Systems

Traditional protein engineering has focused on optimizing individual molecules. However, biology rarely operates in isolation; function is the product of complex interactions between DNA, RNA, and proteins. A key breakthrough in generative genomics is the ability to co-design these interacting components. The Evo model has demonstrated this by successfully generating novel CRISPR-Cas systems [5]. A CRISPR system consists of a Cas protein (nuclease) and a non-coding RNA guide (gRNA) that must fold into a specific structure to bind the protein. Evo, having learned the joint distribution of these components across millions of microbial genomes, can generate a novel Cas protein sequence and, simultaneously, the compatible gRNA scaffold required for it to function [5].

This capability was further validated through the generation of IS200/IS605 transposable elements, mobile genetic units that move within a genome. The AI-generated transposons were synthesized and tested in the lab, proving to be functional: they could successfully excise themselves from DNA and re-insert, demonstrating that the model had mastered the complex "protein-DNA code" governing transposase binding and cleavage [5]. This is a watershed moment, marking the first time a language model has designed a multi-component biological system involving protein-DNA interactions from scratch.

7.2.2 Generating Regulatory Architectures

While protein design (e.g., AlphaFold, ESM) has matured, the design of regulatory DNA, promoters, enhancers, and 5' UTRs remains a frontier. Regulatory elements dictate when, where, and how much a gene is expressed. Generative models like HyenaDNA and Caduceus, with their deep understanding of long-range dependencies, are uniquely positioned to solve this.

Recent work involves "hallucinating" synthetic promoters with specific tissue-specificity profiles. By conditioning a generative model on cell-type-specific chromatin accessibility data (e.g., from scATAC-seq), researchers can generate promoter sequences that drive expression exclusively in target cells (e.g., hepatocytes) while remaining silent in off-target tissues (e.g., cardiomyocytes). This capability is critical for the safety of gene therapies, where off-target expression can lead to toxicity. The Nucleotide Transformer (NT-v2) has shown an exceptional ability to learn "regulatory grammar," outperforming specialized models like Enformer in predicting the impact of regulatory mutations, suggesting it possesses the latent capacity to design these elements de novo [2].

7.2.3 From Sequence to Function to Fitness

The goal of generative biology is the optimization of fitness. In the context of genomic foundation models, "fitness" is often implicitly encoded as the "likelihood" of a sequence. The Nucleotide Transformer study revealed that the model's likelihood scores correlate strongly with pathogenicity; sequences with low probability under the model are often deleterious or pathogenic [2].

This implies a powerful design paradigm: Evolution via Gradient Ascent. Instead of screening random mutations, we can start with a wild-type sequence and iteratively update it to maximize its probability under the model (or a specific condition token). This in silico evolution can optimize sequences for stability, expression level, or binding affinity. For example, optimizing a metabolic pathway in a chassis organism (like E. coli) could involve generating a 50-kb sequence containing multiple genes and regulatory elements, optimized globally for flux through the pathway, a task that is currently impossible for human designers but within the theoretical reach of 1 M-context models like HyenaDNA [5].

7.3 AI-Driven Perturbation Prediction and Digital Twins

One of the "Holy Grails" of computational biology is the creation of a Digital Twin: a virtual replica of a biological system (cell, organ, or patient) that can simulate responses to stimuli. This capability moves biology from an observational science to a predictive, simulation-based engineering discipline.

7.3.1 Single-Cell Foundation Models: The Engine of Virtual Cells

The foundation of the biological digital twin is the single cell. The explosion of single-cell RNA sequencing (scRNA-seq) has provided the data necessary to train "Cellular Large Language Models." Leading architectures in this space include scGPT [7] and Geneformer [8]. scGPT, developed by the Bo Wang Lab, treats a single cell as a "sentence" and genes as "words." It is pre-trained on over 33 million single-cell transcriptomes from the CZ

CELLxGENE Census [7]. A key innovation in scGPT is the use of Condition Tokens and Value Binning.

- **Value Binning**: Gene expression is continuous, not discrete like text. scGPT bins expression values into relative tokens, allowing the Transformer to process expression levels as semantic information [7].
- **Condition Tokens**: The model accepts special tokens representing experimental conditions (e.g., "Drug X treated," "Batch A"). This allows for explicit modeling of perturbations.

Geneformer [8] utilizes a transfer learning approach pre-trained on human tissues to predict gene network dynamics. A landmark application of Geneformer involved modeling cardiomyopathy. In an iPSC-based model of the disease, Geneformer was able to identify candidate therapeutic targets that, when experimentally inhibited, significantly improved cardiomyocyte contraction [8]. Crucially, Geneformer achieved this with limited disease-specific data, leveraging its "general knowledge" of gene regulation learned during pre-training to infer the correct targets. This demonstrates the power of foundation models to bridge the gap between small, specific clinical datasets and massive, general biological knowledge.

7.3.2 Zero-Shot Perturbation Prediction

The most transformative capability of these models is zero-shot perturbation prediction. Traditional methods require experimental data for every perturbation (e.g., "What happens if we knock out Gene A?"). Foundation models can infer these outcomes. By learning the universal manifold of gene co-expression, scGPT can predict the transcriptomic phenotype of a gene knockout it has never seen in training. If the model has learned that Gene A regulates Gene B, and Gene B regulates Gene C, it can infer that knocking out A will impact C [9]. This allows researchers to perform in silico CRISPR screens, testing millions of genetic perturbations or drug interactions computationally to prioritize a small number of candidates for wet-lab validation.

A schematic shown in Fig. 7.2 illustrates the data flow in a Virtual Cell model. (1) Input: Patient-specific scRNA-seq data. (2) Embedding: scGPT/Geneformer projects cells into a latent space. (3) Perturbation: A "digital drug" vector is added to the latent representation. (4) Decoding: The model reconstructs the predicted post-treatment gene expression profile. (5) Validation: Top predictions are tested in organoids.

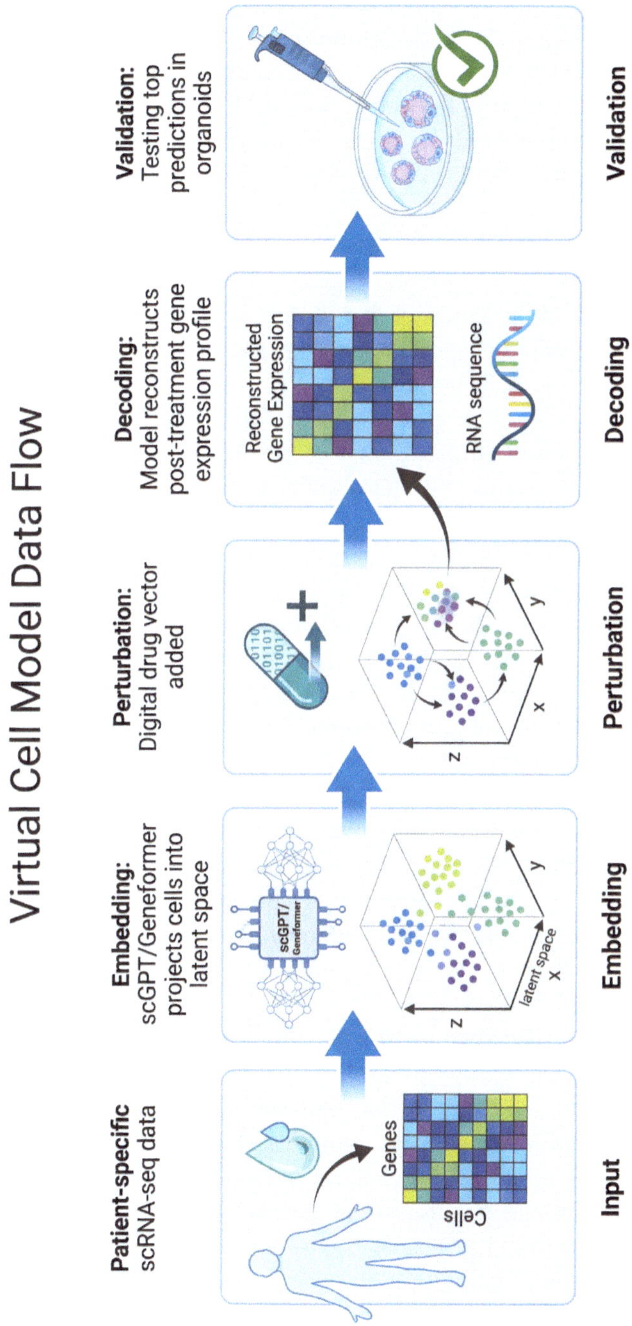

Fig. 7.2 Virtual cell workflow

7.3.3 The Hierarchy of Digital Twins: From Cells to Organs

Moving up the biological hierarchy, we encounter Organ-Level Digital Twins. These models integrate genomic data with physics-based simulations of organ function. A prime example is the Virtual Heart project at Johns Hopkins University (Geno-DT). This initiative creates "genotype-specific digital twins" for patients with Arrhythmogenic Right Ventricular Cardiomyopathy (ARVC) [10]. The model integrates two distinct data streams:

- **Structural**: Contrast-enhanced MRI scans define the patient-specific geometry and fibrosis of the heart.
- **Genomic/Functional**: The patient's specific genetic variants (e.g., in PKP2) are used to parameterize the electrophysiological properties of the virtual cells (e.g., conductivity, restitution) [10].

This virtual heart is then "stressed" in a computer simulation to predict the likelihood of developing ventricular tachycardia (VT). This approach has outperformed standard clinical classifiers, providing a personalized risk assessment that guides decisions on whether to implant a defibrillator [11]. This fusion of mechanistic modeling (physics) with probabilistic modeling (AI/Genomics) represents the future of precision medicine, where the "Digital Twin" is a dynamic, executable medical device.

7.4 Clinical and Phenotypic Integration: Polygenic Risk Scores 2.0 and Multi-omics

As genomic AI transitions into the clinic, the focus shifts from abstract discovery to concrete patient stratification. The traditional Polygenic Risk Score (PRS), a linear summation of common variant effects, is being superseded by "PRS 2.0," which integrates non-linear interactions, rare variants, and multi-omics data layers.

7.4.1 Beyond Linearity: Boosting and Non-coding Interactions

Standard PRS models are limited by their assumption of linearity; they assume that the risk contributed by Variant A is independent of Variant B. Biology, however, is defined by epistasis (gene–gene interactions) and non-linear thresholds. New AI-driven approaches are addressing this by applying non-linear algorithms to Genome-Wide Association Study (GWAS) data. Component-wise L2-boosting (snpboost) is one such method that iteratively selects variants most correlated with the residuals of the previous iteration [12]. This allows the model to capture complex, non-additive architectures of disease risk without the massive computational cost of a full neural network search.

Furthermore, the integration of rare variants is critical. Common variants often explain only a fraction of phenotypic variance ("missing heritability"). Deep learning models are now used to aggregate rare variants particularly in non-coding regions into "pathway burden scores." By using deleterious scores from foundation models (like Nucleotide Transformer or CADD) to weight rare variants and then collapsing them into functional units (genes or enhancers), researchers can detect signal in WGS data that standard GWAS misses [13]. This approach has been particularly fruitful in blood traits and complex diseases where rare, high-effect alleles play a significant role.

7.4.2 TEASE-MM and Tensor-Based Multi-omics

Genomics provides the blueprint, but transcriptomics, proteomics, and metabolomics provide the state. Integrating these heterogeneous data types is statistically challenging. Tensor-based Association Test for Multi-omics (TEASE-MM) introduces a rigorous mathematical framework for this integration. Instead of flattening multi-omics data into a single long vector (concatenation), TEASE-MM represents the data as a Tensor (a multi-dimensional array) [13]. This preserves the structural relationships between subjects, tissues, and omic layers.

- **Mechanism**: TEASE-MM utilizes a tensor-based variance component test that allows for "variable-wise" inference. It can determine not just if a set of omics features is associated with a disease, but which specific modality (e.g., methylation vs. expression) drives that association [14].
- **Application**: Applied to the Cancer Cell Line Encyclopedia (CCLE), TEASE-MM demonstrated superior power in identifying gene sets associated with drug response compared to standard matrix-based methods [14].

7.4.3 Longitudinal Disease Trajectories with Transformers

Disease is a temporal process. Static snapshots fail to capture the velocity and acceleration of pathology. Longitudinal multimodal Transformers are revolutionizing the prediction of disease trajectories by analyzing time-series omics and clinical data [15].

TSCytoPred is a deep learning framework designed to infer continuous cytokine expression trajectories from irregular, sparse longitudinal data [16]. Cytokines are volatile markers of inflammation (e.g., in COVID-19 or Sepsis). TSCytoPred uses a Transformer-based architecture to learn the temporal dependencies between gene expression (which is stable) and cytokine levels (which are volatile), effectively "imputing" the missing time points to reconstruct the full inflammatory trajectory of a patient.

Similarly, Transformer models applied to Electronic Health Records (EHR) and longitudinal biobanks can predict the onset of comorbidities (e.g., Alzheimer's following cardiovascular disease) years in advance [17]. These models learn distinct "health state embeddings" that evolve over time, allowing clinicians to intervene when a patient's trajectory begins to deviate toward a high-risk attractor state.

7.5 Real-Time and Edge Genomics: AI at the Sensor

The centralization of genomics sending samples to a core facility and waiting weeks for data is incompatible with the needs of infectious disease surveillance and acute care. The convergence of Nanopore Sequencing and Edge AI is enabling a new paradigm: "Genomics of Things," where sequencing and analysis happen autonomously at the point of sample collection.

7.5.1 Nanopore Adaptive Sampling: The AI Gatekeeper

Oxford Nanopore Technologies (ONT) sequencers possess a capability unique in the industry: Adaptive Sampling (or "Read Until"). The sequencer can reverse the voltage across individual pores to eject a DNA molecule while it is being sequenced, based on real-time software decisions [18]. AI algorithms act as the gatekeeper in this process.

SquiggleNet utilizes a deep Convolutional Neural Network (CNN) to classify the raw electrical signal (the "squiggle") directly, without base calling [19]. It can distinguish pathogen DNA from host DNA within hundreds of milliseconds. UNCALLED maps the raw signal to a reference using a signal-space algorithm [19].

- **Performance**: In benchmarking studies, these tools have achieved enrichment factors ranging from 1.67 × to 13.87 × for target pathogens in metagenomic backgrounds [20]. This means a low-abundance pathogen (e.g., a virus in blood) can be sequenced to high depth without wasting resources sequencing the human genome. Table 7.2 shows comparison of real-time adaptive sampling algorithms.

7.5.2 Edge Computing on the MinION Mk1C and Mobile Devices

The deployment of these models on Edge Devices enables fully autonomous surveillance. The MinION Mk1C, which integrates a sequencer with an NVIDIA Jetson TX2 GPU, is the standard platform for this [23]. However, the computational demands of Transformer-based basecallers (like Dorado) are pushing the limits of this hardware. Recent innovations have demonstrated the feasibility of running pathogen classifiers on even more

Table 7.2 Comparison of real-time adaptive sampling algorithms

Algorithm	Method	Hardware	Key advantage	Limitation
SquiggleNet [19]	Deep CNN (1D ResNet)	GPU/CPU	Ref-free learning, high speed	Requires training on target squiggles
UNCALLED [19]	Signal mapping (FM-index)	CPU	Accurate mapping	Slower; requires ~ 1.3 kb of read to decide
ReadBouncer [21]	Interleaved bloom filter	CPU	Low memory footprint	Lower sensitivity than deep learning
Dorado [22]	Transformer/ RNN	GPU	Integrated base calling, methylation	High compute requirement (GPU)

constrained hardware, such as Raspberry Pi and smartphones. DeepNano-blitz provides a highly optimized CPU basecaller for low-power devices [19].

- **Smartphone Microscopy**: AI models running locally on smartphones (using TensorFlow Lite) have been used to detect microfilariae (parasites) in blood samples with 99.8% accuracy, demonstrating that high-fidelity diagnostics can function without internet connectivity or cloud servers [24].

This "Edge Genomics" capability is critical for biosecurity and global health. It allows for the detection of drug-resistant tuberculosis or emerging viral threats in remote locations (e.g., rural Africa, mining sites) where data upload to the cloud is impossible [25].

7.6 Autonomous Science and Self-driving Labs

The "Reproducibility Crisis" in biology is, in part, a crisis of manual labor. Human execution of protocols is inherently variable. Self-driving laboratories (SDLs), also known as Cloud Labs, solve this by automating the physical execution of experiments and closing the loop with AI-driven experimental design.

7.6.1 The Closed-Loop Discovery Cycle

An SDL operates on a tight integration of hardware and software, often referred to as the Design-Build-Test-Learn (DBTL) cycle.

- **Design (AI Agent)**: An AI model (e.g., a Bayesian Optimizer or a Foundation Model like Evo) proposes a set of experiments. For example, "Synthesize these 96 protein variants."
- **Build/Test (Robotics)**: Automated liquid handlers and analytical instruments (LC–MS, sequencers) execute the protocol in a facility like Emerald Cloud Lab (ECL) or Strateos [26].
- **Learn (Active Learning)**: The results are structured and fed back to AI. The AI updates its surrogate model of the fitness landscape and selects the next batch of experiments to maximize Expected Improvement (EI) or Upper Confidence Bound (UCB) [27].

7.6.2 Accelerating Discovery: The A-Lab and Beyond

The integration of this loop has led to dramatic accelerations in materials science and is now transforming genomics. The Autonomous Lab (A-Lab) at Lawrence Berkeley National Lab utilized this approach to synthesize novel materials, achieving a success rate of 71–74% by allowing the AI to learn synthesis heuristics from its own failures [28]. In genomics, this approach is being used to traverse the "fitness landscape" of proteins. A historical precursor, Meta-DENDRAL (1970s), used early AI to infer cleavage rules for mass spectrometry [28]. Today, modern SDLs use Active Learning to optimize enzymes. By testing only a tiny fraction of the possible sequence space, the AI "learns" the landscape and zooms in on high-activity variants. This turns scientific discovery from a series of ad-hoc trials into a systematic optimization problem.

7.7 Ethical, Biosecurity, and Governance Frontiers

The capabilities detailed in this chapter generative design of functional biological systems, unmonitored edge sequencing, and autonomous experimentation introduce profound dual-use risks. The same AI that can design a life-saving gene therapy vector can, in principle, design a pathogen with enhanced transmissibility or immune evasion.

7.7.1 The Biosecurity Dilemma of Generative Models

Traditional biosecurity relies on sequence matching. DNA synthesis provides screen orders against a database of known pathogens (e.g., Ebola, Anthrax). If a sequence matches, the order is flagged.

Generative AI renders this safeguard obsolete. Models like Evo or ProteinMPNN can design proteins that are functionally identical to toxins but share little to no sequence homology with known natural counterparts [29]. A "stealth" toxin designed by AI would pass through current screening filters undetected.

- **Red Teaming Studies**: Recent evaluations by RAND and OpenAI assessed the risk of LLMs providing "uplift" to malicious actors planning biological attacks. While current results suggest that today's models do not significantly outperform a standard internet search (Google) in providing access to dual-use information [30], this is a temporary baseline. As models become more specialized in biological engineering (rather than just general knowledge), the risk shifts from "information access" to "functional design capability."

7.7.2 The NIST AI Risk Management Framework (RMF)

To address these emerging threats, the US National Institute of Standards and Technology (NIST) released the Generative AI Profile for the AI Risk Management Framework (AI RMF) in July 2024 [31]. This document establishes the first federal standards for managing the risks of "chem-bio" AI. Key provisions include:

- **Risk Re-categorization**: Moving from monitoring "CBRN Information" to monitoring "CBRN Capabilities" focusing on what the model can do, not just what it "knows" [32].
- **Pre-release Red Teaming**: Mandating that developers of biological foundation models rigorously test for the ability to generate hazardous designs before public release.
- **Know Your Customer (KYC)**: Strengthening verification for access to powerful design tools and cloud labs.

7.7.3 Executive Orders and Global Governance

The regulatory landscape is hardening. The White House Executive Order on the Safe, Secure, and Trustworthy Development and Use of AI (EO 14,110) explicitly targets biosecurity [33]. It mandates that agencies funding life-science research must establish new standards for biological synthesis screening that account for AI-generated sequences. It also calls for the establishment of an AI Safety Institute to oversee the evaluation of frontier models [34]. The future of genomic AI will be defined by this tension between open science and security. We are moving toward a governance model of "secure-by-design," where safety constraints are embedded directly into the loss functions of the models themselves, ensuring that the AI of the future is physically incapable of designing harm.

7.8 Summary

The trajectory of AI in genomics is clear: we are moving from observation to intervention, from correlation to causation, and from automation to autonomy.

- **Architectural Maturity**: The field has moved beyond the naive application of NLP Transformers. Architectures like Hyena and Caduceus now respect the fundamental physics of DNA, its length and its symmetry enabling whole-genome understanding.
- **Generative Power**: We have crossed the threshold from reading biology to writing it. Models like Evo are not just predicting properties; they are co-designing complex, functional molecular systems like CRISPR and transposons.
- **Digital Twins**: The integration of AI with mechanistic modeling (Geno-DT) is creating executable simulations of patients, allowing for in silico clinical trials and personalized risk assessment.
- **The Loop is Closed**: With Edge AI (Nanopore) and self-driving labs, the latency between data acquisition, analysis, and experimental action is collapsing to zero.
- **The Responsibility Gap**: As these capabilities grow, so do the risks. The "dual-use" nature of generative biology demands a new era of proactive governance, red-teaming, and biosecurity-aware engineering.

The next decade of genomic AI will not be defined merely by larger models, but by models that are more physically grounded, biologically plausible, and ethically aligned with the preservation of human health.

References

1. Zhou, Z., Ji, Y., Li, W., Dutta, P., Davuluri, R., & Liu, H. (2023). *Dnabert-2: Efficient foundation model and benchmark for multi-species genome. arXiv preprint* arXiv:2306.15006.
2. Feng, H., Wu, L., Zhao, B., Huff, C., Zhang, J., Wu, J., ... & Wu, C. (2025). Benchmarking DNA foundation models for genomic and genetic tasks. *Nature communications, 16*(1), 10780.
3. Schiff, Y., Kao, C. H., Gokaslan, A., Dao, T., Gu, A., & Kuleshov, V. (2024). Caduceus: Bi-directional equivariant long-range dna sequence modeling. *Proceedings of machine learning research, 235*, 43632.
4. Feng, H., Wu, L., Zhao, B., Huff, C., Zhang, J., Wu, J., ... & Wu, C. (2024). *Benchmarking DNA foundation models for genomic sequence classification. bioRxiv.*
5. Nguyen, E., Poli, M., Durrant, M. G., Kang, B., Katrekar, D., Li, D. B., ... & Hie, B. L. (2024). Sequence modeling and design from molecular to genome scale with Evo. *Science, 386*(6723), eado9336.
6. ArcInstitute/evo2: *Genome modeling and design across all domains of life.* Weblink: https://github.com/ArcInstitute/evo2. Last Access: February 2026.
7. scGPT, v1.0, *Virtual Cells Platform.* Weblink: https://virtualcellmodels.cziscience.com/model/scgpt. Last Access: February 2026.

8. Theodoris, C. V., Xiao, L., Chopra, A., Chaffin, M. D., Al Sayed, Z. R., Hill, M. C., ... & Ellinor, P. T. (2023). Transfer learning enables predictions in network biology. *Nature, 618*(7965), 616–624.
9. Cui, H., Wang, C., Maan, H., Pang, K., Luo, F., Duan, N., & Wang, B. (2024). ScGPT: Toward building a foundation model for single-cell multi-omics using generative AI. *Nature methods, 21*(8), 1470–1480.
10. *'Digital twins' of patients' hearts harness personal genetics to inform disease management.* Weblink: https://www.bme.jhu.edu/news-events/news/digital-twins-of-patients-hearts-harness-personal-genetics-to-inform-disease-management/. Last Access: February 2026.
11. Sel, K., Osman, D., Zare, F., Masoumi Shahrbabak, S., Brattain, L., Hahn, J. O., ... & Jafari, R. (2024). Building digital twins for cardiovascular health: from principles to clinical impact. *Journal of the American Heart Association, 13*(19), e031981.
12. Klinkhammer, H., Staerk, C., Maj, C., Krawitz, P. M., & Mayr, A. (2023). A statistical boosting framework for polygenic risk scores based on large-scale genotype data. *Frontiers in Genetics, 13*, 1076440.
13. Ribeiro, D. M., Hofmeister, R. J., Rubinacci, S., & Delaneau, O. (2025). Noncoding rare variant associations with blood traits in 166,740 UK Biobank genomes. *Nature Genetics, 57*(9), 2146–2155.
14. Chang, S. M., Yang, M., Lu, W., Huang, Y. J., Huang, Y., Hung, H., ... & Tzeng, J. Y. (2021). Gene-set integrative analysis of multi-omics data using tensor-based association test. *Bioinformatics, 37*(16), 2259–2265.
15. Li, T. Z., Still, J. M., Xu, K., Lee, H. H., Cai, L. Y., Krishnan, A. R., ... & Lasko, T. A. (2023, October). Longitudinal multimodal transformer integrating imaging and latent clinical signatures from routine EHRs for pulmonary nodule classification. In *International Conference on Medical Image Computing and Computer-Assisted Intervention* (pp. 649–659). Cham: Springer Nature Switzerland.
16. Choi, J. M., & Chae, H. (2025). TSCytoPred: A deep learning framework for inferring cytokine expression trajectories from irregular longitudinal gene expression data to enhance multi-omics analyses. *PeerJ, 13*, Article e20270.
17. Taheriyoun, A. R., Ross, A., Safikhani, A., Soudbakhsh, D., & Rahnavard, A. (2026). Longitudinal omics data analysis: approaches and applications. *Computational and Structural Biotechnology Journal.*
18. Adaptive sampling – Oxford Nanopore Technologies. Weblink: https://nanoporetech.com/document/adaptive-sampling. Last Access: February 2026.
19. Yang, L., Lin, Y., Li, P., Wang, K., Li, J., Liu, Y., ... & Song, H. (2025). A comprehensive benchmarking of adaptive sampling tools for nanopore sequencing. *Genome Biology, 26*(1), 281.
20. Martin, S., Heavens, D., Lan, Y., Horsfield, S., Clark, M. D., & Leggett, R. M. (2022). Nanopore adaptive sampling: A tool for enrichment of low abundance species in metagenomic samples. *Genome biology, 23*(1), 11.
21. Ulrich, J. U., Lutfi, A., Rutzen, K., & Renard, B. Y. (2022). ReadBouncer: precise and scalable adaptive sampling for nanopore sequencing. *Bioinformatics, 38*(Supplement_1), i153-i160.
22. Data Analysis—Oxford Nanopore Technologies. Weblink: https://nanoporetech.com/document/data-analysis. Last Access: February 2026.
23. Simon, W. A., Boybat, I., Kodra, R., Ferro, E., Singh, G., Alser, M., ... & Sebastian, A. (2025). Cimba: Accelerating genome sequencing through on-device basecalling via compute-in-memory. *IEEE Transactions on Parallel and Distributed Systems.*

24. Lin, L., Dacal, E., Díez, N., Carmona, C., Martin Ramirez, A., Barón Argos, L., ... & Luengo-Oroz, M. (2024). Edge Artificial Intelligence (AI) for real-time automatic quantification of filariasis in mobile microscopy. *PLOS Neglected Tropical Diseases, 18*(4), e0012117.
25. Humphreys, G. (2026). *Frontline genomics with AI: Nanopores, lifesaving diagnostics, and squiggles.* Weblink: https://globalhealthnow.org/2026-01/frontline-genomics-ai-nanopores-lifesaving-diagnostics-and-squiggles. Last Access: February 2026.
26. Emerald Cloud Lab: Remote Controlled Life Sciences Lab. Weblink: https://www.emeraldcloud lab.com/. Last Access: February 2026.
27. Tom, G., Schmid, S. P., Baird, S. G., Cao, Y., Darvish, K., Hao, H., ... & Aspuru-Guzik, A. (2024). Self-driving laboratories for chemistry and materials science. *Chemical Reviews, 124*(16), 9633–9732.
28. Tobias, A. V., & Wahab, A. (2025). Autonomous 'self-driving'laboratories: A review of technology and policy implications. *Royal Society Open Science, 12*(7), Article 250646.
29. Closing the Biosecurity Gap in Synthetic Biology—Global Biodefense. Weblink: https://glo balbiodefense.com/2025/10/07/closing-the-biosecurity-gap-in-synthetic-biology/. Last Access: February 2026.
30. Red-Teaming the Risks of Using AI in Biological Attacks. Weblink: https://www.rand.org/pubs/articles/2024/red-teaming-the-risks-of-using-ai-in-biological-attacks.html. Last Access: February 2026.
31. AI Risk Management Framework—National Institute of Standards and Technology. Weblink: https://www.nist.gov/itl/ai-risk-management-framework. Last Access: February 2026.
32. NIST AI 6001. *Artificial Intelligence Risk Management Framework—Johns Hopkins Center for Health Security*. Weblink: https://centerforhealthsecurity.org/sites/default/files/2024-06/2024-06-02-jhchs-nist-ai-6001-rfc.pdf. Last Access: February 2026.
33. Improving the Safety and Security of Biological Research—The White House. Weblink: https://www.whitehouse.gov/presidential-actions/2025/05/improving-the-safety-and-security-of-bio logical-research/. Last Access: February 2026.
34. White House Launches AI Action Plan and Executive Orders to Promote Innovation, Infrastructure, and International Diplomacy and Security. Weblink: https://www.wiley.law/alert-White-House-Launches-AI-Action-Plan-and-Executive-Orders-to-Promote-Innovation-Infras tructure-and-International-Diplomacy-and-Security. Last Access: February 2026.

Index

K. Shaikh et al., *Artificial Intelligence in Genomics*, Synthesis Lectures on Engineering, Science, and Technology, https://doi.org/10.1007/978-3-032-25486-3

P

Q

R

S

T

V

W

Z

MIX
Papier aus verantwortungsvollen Quellen
Paper from responsible sources
FSC® C105338

If you have any concerns about our products,
you can contact us on
ProductSafety@springernature.com

In case Publisher is established outside the EU,
the EU authorized representative is:
Springer Nature Customer Service Center GmbH
Europaplatz 3, 69115 Heidelberg, Germany

Printed by Libri Plureos GmbH
in Hamburg, Germany